Antonio M. Gotto, Jr (Ed.)

Cellular and Molecular Biology of Atherosclerosis

With 37 Figures

Springer-Verlag

London   Berlin   Heidelberg   New York
Paris   Tokyo   Hong Kong
Barcelona   Budapest

Antonio M. Gotto, Jr, MD, DPhil
The Methodist Hospital and Baylor College of Medicine,
Houston, Texas 77030, USA

Cover illustrations: Ch. 1, Fig. 2. Segment of a coronary artery which is entirely
normal. (Reproduced by permission from Arnett et al. 1979.)
Ch. 7, Fig. 1. Increased leukocyte adherence to the aortic endothelial surface after
MM-LDL injection.

ISBN 3-540-19704-4 Springer-Verlag Berlin Heidelberg New York
ISBN 0-387-19704-4 Springer-Verlag New York Berlin Heidelberg

British Library Cataloguing in Publication Data
Cellular and molecular biology of atherosclerosis.
  I. Gotto, Antonio M.
  616.136
ISBN 3540197044

Library of Congress Cataloging-in-Publication Data
Cellular and molecular biology of atherosclerosis/Antonio M. Gotto.
  Jr. (ed.).
      p. cm.
  Proceedings of a symposium held Oct. 29–30, 1990 in Brussels,
Belgium and sponsored by the Princesse Liliane Cardiology
Foundation.
  Includes index.
  ISBN 3-540-19704-4. — ISBN 0-387-19704-4
  1. Atherosclerosis—Molecular aspects—Congresses.
2. Atherosclerosis—Cytopathology—Congresses. I. Gotto, Antonio
M. II. Foundation cardiologique Princesse Liliane.
  [DNLM: 1. Atherosclerosis—etiology—congresses.   WG 550 C393
1990]
RC692.C44   1992
616.1'3607—dc20
DNLM/DLC   91–5145
for Library of Congress   CIP

Composition by Genesis Typesetting, Laser Quay, Rochester, Kent, UK
Printed by Page Bros, Norwich. Bound by the Bath Press, Bath
12/3830-543210   Printed on acid-free paper

Preface

Atherosclerotic cardiovascular disease remains the major cause of death and disability in Western society. The field of atherosclerosis research has grown tremendously over the last forty years, shedding a great deal of light on the contributing factors and natural history of the disorder and enabling strategies for its treatment and prevention. Some of the greatest strides in this field in recent years have derived from advances in molecular biology techniques. These strides were chosen for emphasis in the most recent Princess Lilian symposium, whose proceedings this volume represents.

Historically, the Princess Lilian meetings have been small ones aimed at bringing together investigators from diverse specialties to discuss a particular subject. The most recent meeting was no exception and included clinicians, clinical investigators, and researchers in basic science.

The symposium began with an extensive review of coronary morphopathological findings in patients who died of coronary heart disease. Any rational hypothesis of atherogenesis must take into account clinical findings, and any attempt to bridge the gap between experimental laboratory findings and studies in man is highly desirable.

Three chapters focus on endothelial injury: one on the nitric oxide pathway in physiology and pathology, a second on the activation of endothelial cells, and a third on the monocyte and endothelial injury. Still another chapter examines growth factors, in particular the fibroblast growth factor in atherogenesis.

Much interest has recently focused on the possible role of oxidized and modified lipoproteins in atherogenesis. Clinical trials are being planned to test whether the administration of antioxidants will protect against atherosclerosis. The roles of modified lipoproteins are reviewed in this volume, as is the possible contribution of lipoprotein [a], an independent predictor of coronary disease. The roles of apolipoproteins are also discussed, from the molecular biology of apolipoprotein B (the most atherogenic of the apolipoproteins), to

apolipoprotein mutations, to the regulation of lipoprotein metabolism through receptor mechanisms. Recent evidence has suggested
that the presence of a large fraction of small, dense low-density
lipoprotein particles in conjunction with elevations of triglycerides
and low levels of high-density lipoprotein predisposes to coronary
disease. One mutation that could produce this disorder is the
heterozygous state for lipoprotein lipase deficiency. Postprandial
lipemia, low levels of high-density lipoprotein, and hypertriglyceridemia are discussed from the points of view of basic research and
clinical manifestations.

No specific marker of apolipoproteins using the restriction
fragment-length polymorphism (RFLP) technique has yet been
shown to be predictive in coronary disease in population groups.
However, a number of interesting observations have been made from
RFLP studies regarding apolipoprotien mutations, which was the
subject of one symposium presentation.

Another chapter examines the etiology of hypertension in a strain
of genetically hypertensive rats. The way in which hypertension and
hyperlipidemia interact to induce atherosclerosis is not known.

The volume closes with a current and future perspective on clinical
trials with lipid-lowering agents.

The meeting from which this book was derived also fostered much
productive discussion not included in this volume, since participants
were able to interact informally as well to discuss their findings. This
opportunity – within a spectacular program of social activities – was
made possible through the generosity of Princess Lilian of Belgium
and the Princess Lilian Cardiology Foundation.

July 1991   *Antonio M. Gotto, Jr, MD, DPhil*

Contents

11   Genetic Control of Plasma Lipid, Lipoprotein and Apolipoprotein Levels: From Restriction Fragment Length Polymorphisms to Specific Mutations
S. Humphries, A. Dunning, Chun-Fang Xu and P. Talmud

12   Receptor Regulation of Lipoprotein Metabolism
U. Beisiegel

13   The High-Density Lipoprotein Receptor
J. F. Oram

List of Main Contributors

G. Assmann
Institut für Klinische Chemie und Laboratoriumsmedizin
Albert-Schweitzer-Strasse 33
D-4400 Munster
Germany

F. E. Baralle
International Centre for Genetic Engineering and Biotechnology
Padriciano 99
34012 Trieste
Italy

U. Beisiegel
I. Medizinische Klinik
Biochemisches Stoffwechsellabor
Universität Hamburg
Universität-Krankenhaus Eppendorf
Martinstrasse 52
2000 Hamburg 20-07
Germany

J. Berliner
Department of Pathology
UCLA School of Medicine
Center for Health Sciences
10822 Le Conte Avenue
Los Angeles, California 90024-1732
USA

W. Casscells
Department of Molecular and Cellular Growth Biology
The Whittier Institute for Diabetes and Endocrinology at
Scripps Institutes of Medicine and Science
9894 Genesee Avenue
La Jolla, CA 92037
USA

L. Chan
Departments of Cell Biology and Medicine
Baylor College of Medicine
One Baylor Plaza
Houston, Texas 77030
USA

R. Cotran
Department of Pathology
Brigham and Women's Hospital
75 Francis Street
Boston, Massachusetts 02115
USA

R. Gerrity
School of Medicine
Department of Pathology
Section of Anatomic Pathology
The Medical College of Georgia
Augusta, Georgia 30912-3605
USA

A. Gotto
Department of Internal Medicine
The Methodist Hospital
6565 Fannin, M.S. SM 1423
Houston, Texas 77030
USA

S. Humphries
Arterial Disease Research Unit
Charing Cross Sunley Research Centre
Lorgan Avenue
Hammersmith
London, W6 8LW
UK

S. Moncada
Wellcome Research Laboratories
Langley Court
Beckenham
Kent BR3 3BS
UK

J. Oram
Department of Medicine, RG-26
Division of Metabolism, Endocrinology & Nutrition
University of Washington
Seattle, Washington 98195
USA

J. R. Patsch
Division of Clinical Atherosclerosis Research
Department of Medicine
University of Innsbruck
A-6020
Austria

W. Roberts
National Heart, Lung and Blood Institute
Building 10/Room 2N258
National Institute of Health
9000 Rockville Pike
Bethesda, Maryland 20892
USA

A Scanu
Department of Medicine
University of Chicago
5841 South Maryland (BOS 231)
Chicago, Illinois 60637
USA

J. Scott
Clinical Research Centre
Division of Molecular Medicine
Watford Road
Harrow
Middlesex HA1 3UJ
UK

Chapter 1

Morphological Findings in the Coronary Arteries in Fatal Coronary Artery Disease

W. C. Roberts

Introduction

Atherosclerotic coronary artery disease (CAD) is the most common cause of death in the Western world. One American dies every minute because of atherosclerotic CAD. In the USA alone, about 6 million persons have symptomatic myocardial ischemia because of atherosclerotic CAD. About 250 000 coronary artery bypass grafting operations were performed in 1990 in the USA and about 300 000 coronary angioplasty procedures. The cause of atherosclerosis is now clear. The evidence is overwhelming that atherosclerosis is a cholesterol problem. The higher the blood total cholesterol level (specifically the low-density lipoprotein level) the greater the chance of developing symptomatic CAD, the greater the chance of having fatal CAD, and the greater the extent of the atherosclerotic plaques. Furthermore, lowering the blood total cholesterol level decreases the chances of having symptomatic or fatal CAD and the greater the chance that some atherosclerotic plaques will actually become smaller, i.e. regress. Although the coronary arteries have been examined by visual inspection at necropsy for over 100 years, only in recent years has the extent of the atherosclerotic process in the coronary arteries in patients with symptomatic or fatal CAD become appreciated. This chapter initially reviews the status of the major epicardial coronary arteries in various subsets of patients with fatal atherosclerotic CAD. It then describes the effects of angioplasty on these arteries, some observations in patients having thrombolytic therapy and coronary bypass, and then various complications of myocardial ischemia.

Number of Major Epicardial Coronary Arteries Severely Narrowed in the Various "Coronary Events"

The most common method for describing the severity of CAD in patients with clinical evidence of myocardial ischemia is by the number of major epicardial coronary arteries narrowed >50% in luminal diameter by angiogram. Thus, patients are divided into groups of 1-vessel, 2-vessel, 3-vessel and "left main" CAD. Because a 50% diameter reduction in general is equivalent to a 75% cross-sectional area narrowing, the cut-off point of "significant" as opposed to "insignificant" luminal narrowing at necropsy is the 75% cross-sectional area point. Physiologically, there is no obstruction to arterial flow until the lumen is narrowed >75% in cross-sectional area.

Table 1.1 summarizes the number of major (right, left main, left anterior descending, and left circumflex) epicardial coronary arteries narrowed >75% in cross-sectional area by atherosclerotic plaque alone in patients with fatal CAD (Roberts 1989). Among the 129 patients with fatal CAD studied at necropsy, 516 major epicardial coronary arteries were examined and of them 345 (67%) were narrowed at some point 76%–100% in cross-sectional area by atherosclerotic plaque. In contrast, of 40 control subjects, mainly victims of acute leukemia, and without clinical evidence of myocardial ischemia during life, 160 major epicardial coronary arteries were examined and of them 60 (37%) were narrowed at some point >75% in cross-sectional area by plaque. Among the 129 coronary patients, only 11 (8%) had a single coronary artery severely narrowed (controls = 23%); 37 (29%) had two arteries so narrowed (controls = 13%); 64 (50%) had three arteries severely narrowed (controls = 5%) and 17 patients (13%) had all four major arteries so narrowed (controls = 0). Thus, of the four major coronary arteries in the coronary patients an average of 2.7 were narrowed >75% in cross-sectional area by plaque, and among the control subjects 0.7 of four.

Table 1.1. Number of major (right, left main, left anterior descending and left circumflex) coronary arteries narrowed >75% in cross-sectional area by atherosclerotic plaque in fatal coronary artery disease

Coronary event	Pts (n)	Mean age (yrs)	No. of four arteries/Pt > 75% ↓ in CSA by plaque				
			4	3	2	1	Mean
Sudden coronary death	31	47	3	20	6	2	2.8
Acute myocardial infarction	27	59	3	14	10	0	2.7
Healed myocardial infarction							
Asymptomatic	18	66	0	7	7	4	2.2
Chronic CHF without aneurysm	9	63	0	3	5	1	2.2
Left ventricular aneurysm	22	61	1	12	6	3	2.5
Angina pectoris/unstable	22	48	10	8	3	1	3.2
Total (%)	129	56	17 (13)	64 (50)	37 (29)	11 (8)	2.7
Controls (%)	40	52	0 (0)	5 (5)	12 (13)	21 (23)	0.7

CHF = congestive heart failure; CSA = cross-sectional area

The numbers of major coronary arteries severely narrowed by atherosclerotic plaque among the various subsets of coronary patients was relatively similar except for the unstable angina patients (Table 1.1). Among the 31 *sudden coronary death* patients (Roberts and Jones 1979) all of whom died outside the hospital, usually within a few minutes of onset of symptoms of myocardial ischemia, an average of 2.8 of the four major arteries were severely narrowed, a number virtually identical to that of the 27 patients with *transmural acute myocardial infarction* (Roberts and Jones 1980), all of whom died in a coronary care unit. Only two of the 31 sudden death victims and none of the 27 acute myocardial infarction victims had only a single coronary artery ("1-vessel disease") severely narrowed.

The *healed myocardial infarction* group was divided into three subgroups. One consisted of patients who had had an acute myocardial infarction in the past and it had healed and thereafter there was never evidence of myocardial ischemia clinically, and these patients died from a non-cardiac cause, usually cancer (Virmani and Roberts 1981). Nevertheless, the average number of major coronary arteries severely narrowed at necropsy was 2.2 of four. Another subgroup consisted of patients who had had chronic congestive heart failure after healing of an acute myocardial infarction but in the absence of a left ventricular aneurysm (Virmani and Roberts 1980). This group might be called *ischemic cardiomyopathy*. The average number of major coronary arteries severely narrowed in them also was 2.2 of four. The other subgroup of healed myocardial infarction patients had a true left ventricular aneurysm (Cabin and Roberts 1980). The average number of major coronary arteries severely narrowed in them was 2.5 of four.

The final subgroup consisted of 22 patients with *unstable angina pectoris* and all of them had had coronary artery bypass grafting procedures within seven days of death (Roberts and Virmani 1979). Preoperatively, all had normal left ventricular function and none had had a clinically apparent acute myocardial infarct or congestive heart failure at any time. The average number of major coronary arteries severely narrowed by plaque was 3.2 of four, and 10 of the 22 patients had severe narrowing of the left main coronary artery as well as severe narrowing of the other three major coronary arteries ("4-vessel disease"). (Another study (Bulkley and Roberts 1976) has indicated that severe narrowing of the left main coronary artery usually is an indicator that the other three major arteries also are severely narrowed.) The unstable angina group thus had the largest average number of major coronary arteries severely narrowed of any of the subgroups but nevertheless this group of patients had excellent left ventricular function.

Quantitative Approach to Atherosclerotic Coronary Artery Disease: Amounts of Narrowing in Each 5-mm Segment of Each of the Four Major Coronary Arteries

Although the 1-, 2-, 3-, and 4-vessel disease approach has been useful clinically, this type of severity analysis might be thought of as a *qualitative* approach, and differences in degrees of coronary narrowing in the various subsets of coronary patients is usually not discernible by this approach. To obtain a better appreciation of the extent of the atherosclerotic process in patients with fatal CAD, several years ago my colleagues and I began examining each 5-mm long segment of each of the four major coronary arteries (Roberts 1989). In adults, the average length of the right coronary artery is 10 cm; the left main, 1 cm; the left anterior descending, 10 cm, and the left circumflex, 6 cm. Thus, 27 cm of major epicardial coronary artery are available for examination in each adult. Because each 1 cm is divided into two 5-mm long segments, an average of 54 5-mm segments is available to examine in each heart. This approach not only allows one to ask how many of the 5-mm segments are narrowed 76%–100% in cross-sectional area, but also how many are narrowed 51%–75%, 26%–50%, and 0%–25%. This approach, in contrast to the 1-, 2-, 3-, 4-vessel disease approach, might be considered a *quantitative* one.

The same patients previously described by the qualitative approach also were examined at necropsy by the quantitative approach and the findings are summarized in Table 1.2. A total of 6461 5-mm segments were sectioned and later examined histologically. The sections were stained by the Movat method to delineate the internal elastic membrane. The findings in the 129 coronary patients were compared to those in 1849 5-mm segments in 40 control subjects. In each coronary subgroup the 5-mm segments from each of the four major coronary arteries were pooled together so by this approach the amount of narrowing in an individual patient was not discernible. The percentage of 5-mm segments narrowed 76%–100% in cross-sectional area by atherosclerotic plaque was 35% for the coronary patients and 3% for the control subjects; the percentage narrowed 51%–75% was 36% for the coronary patients and 22% for the control subjects. Thus, 71% of the 5-mm segments in the coronary patients were narrowed >50% in cross-sectional area by atherosclerotic plaque and 25% in the control subjects. In contrast, only 29% of the 5-mm segments in the coronary patients were narrowed <50% and only 8% even approached normal, i.e. narrowed 25% or less in cross-sectional area. In contrast, 75% of the 5-mm segments in the control subjects were narrowed <50% and 31% of them were normal or nearly normal. Thus, in the coronary patients 92% of the 6461 5-mm segments of the four major epicardial coronary arteries were narrowed >25% in cross-sectional area by atherosclerotic plaque. Accordingly, the coronary atherosclerotic process is a diffuse one in patients with fatal CAD. To believe that the atherosclerotic process is a focal one in patients with fatal CAD is to believe a myth.

Among the various subsets of coronary patients, those with *sudden coronary death* (Roberts and Jones 1979) and *acute myocardial infarction* (Roberts and Jones 1980) had similar percentages of 5-mm segments narrowed 76%–100% in cross-sectional area by plaque (36% and 34%, respectively); patients with *healed myocardial infarction* (Virmani and Roberts 1980, 1981; Cabin and Roberts 1980) as a group had the least severe narrowing (31% of segments narrowed >75%), and the patients with *unstable angina pectoris* (Roberts and Virmani 1979) had the most severe narrowing of all (48% of the 5-mm segments were narrowed >75% by plaque).

In an attempt to provide a single number for the amount of coronary arterial narrowing in each patient, a score system was utilized. A segment narrowed 0%–25% was assigned a score of 1; a segment narrowed 26%–50%, 2; a segment narrowed 51%–75%, 3; and one narrowed 76%–100%, a score of 4.

Table 1.2. Amounts of cross-sectional area narrowing of each 5-mm segment of the four major (right, left main, left anterior descending and left circumflex) epicardial coronary arteries by atherosclerotic plaques in subjects with fatal coronary artery disease

Subgroup	Pts (n)	Mean age (yrs)	No. 5-mm segments	Percentage of segments narrowed				Mean score	Mean % narrowing/ 5-mm segments
				0–25	26–50	51–75	76–100		
Sudden coronary death	31	47	1564	7	23	34	36	2.98	67
Acute myocardial infarction	27	59	1403	5	23	38	34	3.01	68
Healed myocardial infarction									
Asymptomatic	18	66	924	11	23	35	31 ⎫	2.87	64
Chronic CHF without aneurysm	9	63	529	11	23	37	29 ⎬ 31%	2.78	61
LV aneurysm	22	61	992	4	21	42	33 ⎭	3.03	68
Angina pectoris	22	48	1049	11	12	29	48	3.12	70
Total	129	56	6461	8	21	36	35	2.98	67
Controls	40	52	1849	31	44	22	3	1.97	32

CHF = congestive heart failure; LV = left ventricular

The mean score for all 129 patients or for each of the 6461 5-mm coronary segments was 3.0 and that for the 40 control subjects or 1849 5-mm segments, 2.0. Again, the unstable angina patients had the most extensive coronary narrowing by this approach.

A possible criticism of the 5-mm segment approach to quantifying coronary arterial narrowing is that the epicardial coronary arteries were fixed in an unphysiological pressure state, namely a zero pressure state, rather than at a systemic arterial diastolic pressure. In an attempt to take into account the unphysiological fixation state, the degrees of narrowing were conservatively judged, i.e. if a segment was more or less in between two quadrants (51%–75% and 76%–100%), always the lesser degree of narrowing was chosen. Secondly, any segment in which a portion of wall was collapsed by the fixation process, the degree of narrowing was determined as if the segment was expanded. And most important, the segments narrowed the most, i.e. >75% in cross-sectional area, were affected the least by the fixation process. Irrespective of whether or not the histological technique employed in this study is perfect or imperfect, the same technique was employed in all subsets of coronary patients and also on all control subjects, and, therefore, the comparison data are highly reliable. Irrespective of whether or not the degrees of luminal narrowing should be slightly greater or slightly less than that determined by this technique, it is clear that the atherosclerotic process is a diffuse one in nearly all patients with fatal CAD. The accuracy of the technique of determining degrees of cross-sectional area narrowing by estimating from stained histological sections magnified about 40 times is similar ($\leq$5%) to that determined by planimetry.

Another possible concern of the aforementioned quantitative data is their applicability to living patients with symptomatic or other clinical evidence (positive exercise test, for example) of myocardial ischemia. It is my view that the major difference in coronary arterial narrowing occurs at the stage of conversion from the asymptomatic to the symptomatic myocardial ischemia state and that there is relatively little difference in degrees of coronary narrowing between the symptomatic and the fatal states. Support for this view can be obtained by the presence of severe and extensive coronary narrowing by angiogram during life, and by studying the coronary tree at necropsy in patients who during life had a coronary event and who died later from a non-cardiac cause. Although data from the latter situation are minimal, the degrees of coronary narrowing at necropsy are similar to that in other patients with symptomatic myocardial ischemia which is fatal. And finally, among the subsets of coronary patients described herein, those with unstable angina pectoris had by far the worst degrees of coronary narrowing and the subjects in this group were the only ones in whom their natural course was interrupted by an iatrogenic event, namely coronary artery bypass grafting (within three days of death).

Distribution of Severe Narrowing in Each of the Three Longest Epicardial Coronary Arteries in Fatal Coronary Artery Disease

In all of the aforementioned quantitative coronary arterial data studied, the amount of cross-sectional area luminal narrowing by atherosclerotic plaque in the right, left anterior descending, and left circumflex coronary arteries was similar if the 5-mm segments in each of the three longest coronary arteries were pooled together from a number of patients. This statement might be best understood by examining a single subset of coronary patients with fatal CAD. Among the 27 patients with fatal transmural acute myocardial infarction, a total of 1358 5-mm long segments were analyzed from the right, left anterior descending and left circumflex coronary arteries, and the percentage of segments narrowed 0%–25%, 26%–50%, 51%–75% and 76%–100% was similar at each of these four categories of cross-sectional area narrowing in each of these three major epidcardial coronary arteries. The same findings were observed in the patients with sudden coronary death, healed myocardial infarction, and unstable angina pectoris (Roberts 1989).

In a single patient, however, the percentage of 5-mm long coronary segments severely (>75% in cross-sectional area) narrowed by atherosclerotic plaque in one major epicardial coronary artery may be greater or lesser than that of another major coronary artery. If the segments from one coronary artery (right, for example), however, were pooled together from several patients with fatal CAD and compared to pooled 5-mm segments from another coronary artery (left anterior descending, for example) from several patients with fatal CAD, the percentage of segments narrowed at each of the four categories of cross-sectional area narrowing in each artery are similar. The definition of "several" has not yet been established, but with few exceptions this principle may apply to as few as three patients with pooled 5-mm segments from each of the three major coronary arteries. Thus, the quantity of atherosclerotic plaque is similar for similar lengths of the right, left anterior descending and left circumflex coronary arteries, and because the amount of atherosclerotic plaque is similar the amount of resulting luminal narrowing also is similar. The cholesterol thesis might not be tenable if the amount of atherosclerotic plaque was highly different in the different major epicardial coronary arteries because the same serum cholesterol level presumably is present in each major coronary artery.

Clinical Usefulness of the Quantitative Approach to Coronary Artery Disease

The information derived at necropsy quantitating the severity and extent of atherosclerosis in the four major epicardial coronary arteries in fatal CAD is potentially useful clinically in two areas: (1) in interpreting degrees of coronary narrowing by angiography during life, and (2) in deciding which of the major coronary arteries needs a conduit at the time of coronary artery bypass grafting.

Without coronary angiography neither coronary bypass nor angioplasty would be done. The only way during life to obtain information on the status of the epicardial coronary arteries is angiography, and, therefore, this procedure revolutionized diagnosis of CAD just as aorto-coronary bypass grafting revolutionized therapy of CAD. But angiography – as good as it is – has certain deficiences. Angiography is a luminogram and a narrowed segment is compared to a less narrowed segment which is assumed to be normal. The angiogram does not delineate the internal elastic membrane of the artery, and, therefore, the artery's true lumen remains uncertain (Figs 1.1 and 1.2).

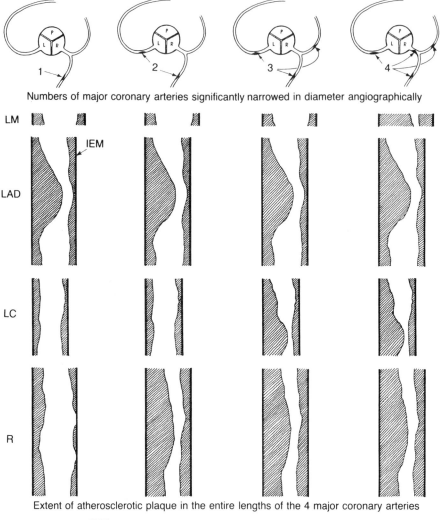

Numbers of major coronary arteries significantly narrowed in diameter angiographically

Extent of atherosclerotic plaque in the entire lengths of the 4 major coronary arteries

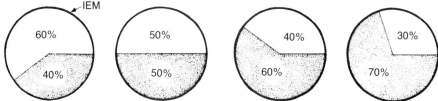

Mean percent of 5-mm segments of all 4 major coronary arteries
narrowed in cross-sectional area by atherosclerotic plaque

Fig. 1.1. Diagram showing numbers of coronary arteries significantly narrowed by angiogram (top), the amounts of plaque in the entire lengths of the four major (left main (LM), left anterior descending (LAD), left circumflex (LC), and right (R)) epicardial coronary arteries (middle), and a cross-sectional area view (bottom) of the average amount of plaque in each 5-mm segment of the entire lengths of the four major coronary arteries according to the numbers of major arteries significantly narrowed by angiogram. (Reproduced with permission from Roberts (1990) Coronary "lesion", coronary "disease", "single-vessel disease", "two-vessel disease": word and phrase misnomers providing false impressions of the extent of coronary atherosclerosis in symptomatic myocardial ischemia. *Am. J. Cardiol.* 66:121–123, 1990.)

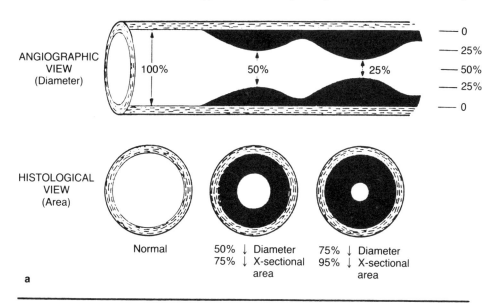

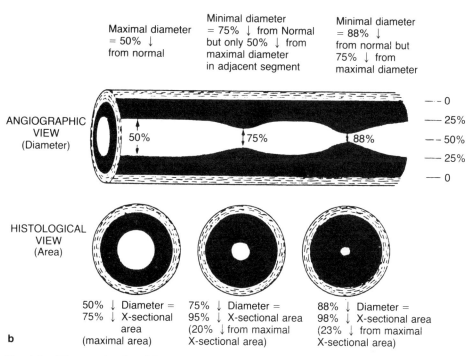

Fig. 1.2. Diagram showing differences in units used for designating degrees of narrowing by angiography (diameter reduction) and that used for histological examination (cross-sectional area narrowing) of coronary arteries. In general, a 50% diameter reduction (by angiography) is equivalent to a 75% cross-sectional area narrowing. The upper panel (**a**) shows a segment of a coronary artery which is entirely normal. This situation rarely exists in persons with symptomatic myocardial ischemia. The usual situation in this is depicted in the lower panel (**b**) which indicates that the severist narrowing is compared to an adjacent area which by angiography may be considered to be normal but in actuality is simply less narrowed. (Reproduced with permission from Arnett et al. 1979.)

The aforementioned coronary quantitative studies demonstrated in fatal CAD that 93% of the 5-mm long segments of the four major epicardial coronary arteries were narrowed >25% in cross-sectional area by atherosclerotic plaque. Thus, only 7% of the 5 mm segments even approached normal and virtually none was normal. Thus, at least in fatal CAD, and probably also in live patients with symptomatic myocardial ischemia, it is infrequent that an angiographically severely narrowed segment of a coronary artery can be compared to a segment of coronary artery which is actually normal. In other words, in patients with symptomatic myocardial ischemia, the coronary angiogram measures degrees of narrowing by comparing severely narrowed segments to segments which are simply less narrowed and by no means normal (Fig. 1.1). Accordingly, coronary angiograms in patients with symptomatic myocardial ischemia usually under-estimate the degrees of luminal narrowing (Arnett et al. 1979; Isner et al. 1981).

The unit of measuring degrees of narrowing by angiography is different from the unit of measurement at necropsy. In the anatomical quantitative studies presented herein, the unit was *cross-sectional area* narrowing. The unit of angiography is *diameter* narrowing. In general, a 75% cross-sectional area narrowing is equivalent to a 50% diameter reduction, and, therefore, a 50% or more diameter reduction during life has generally been considered the cut-off point between clinically significant and clinically insignificant coronary narrowing.

The second potential usefulness of the information derived from the quantitative CAD studies at necropsy is the appreciation that the atherosclerotic process in patients with symptomatic myocardial ischemia is usually diffuse and severe, and, therefore, that more rather than fewer aorto-coronary conduits provide a higher frequency of relief or improvement in symptoms of myocardial ischemia, in improvement in results of exercise testing, and in prolonging life. Among patients surviving <30 days or at later periods after aorto-coronary bypass operations, the amount of severe narrowing in the *non-bypassed native* coronary arteries is usually similar to that in the *bypassed native* coronary arteries (Waller and Roberts 1980; Kalan and Roberts 1990). From study at necropsy of 102 patients dying either early (≤60 days) or late (2.5–108 months (mean 35)) after bypass operations, Waller and Roberts (1980) found that the bypassed and non-bypassed native coronary arteries had similar degrees of severe luminal narrowing by atherosclerotic plaques. Specifically, in 213 (94%) of the 226 bypassed native arteries and in 73 (91%) of 80 non-bypassed native arteries the lumina were narrowed >75% in cross-sectional area by atherosclerotic plaque. The reason the native arteries were not bypassed was not that they were too small or severely narrowed distally, but because by angiogram the lumina were judged not to be sufficiently narrowed to warrant the insertion of a conduit. Thus, if two of the major coronary arteries are severely narrowed by angiogram and the third major artery is "insignificantly" narrowed and if a bypass operation is to be done, the insertion of a conduit in all three major coronary arteries could be reasonably argued. There is, of course, potential danger to inserting a conduit in an artery insignificantly narrowed, but, nevertheless, it may be more advantageous to err on the side of two many conduits than too few. At necropsy, "3-vessel disease" is far more frequent than "2-vessel disease" and even when only two of the three major arteries at necropsy are narrowed >75% in cross-sectional area, the third one is usually narrowed 51%–75% in cross-sectional area. Thus, an appreciation of the diffuse nature of coronary atherosclerosis in fatal CAD and probably also in symptomatic myocardial ischemia encourages the tilt towards more rather than fewer conduits at coronary bypass operations.

Significance of Coronary Arterial Thrombus in Transmural Acute Myocardial Infarction

Thrombi in the coronary arteries of necropsy patients with transmural acute myocardial infarction have been observed in numerous studies (Wartman and Hellerstein 1948; Roberts and Buja 1972; Kragel et al. 1991). Herrick (1919) found them in four patients with fatal acute myocardial infarction, and for many decades thrombi were believed to have precipitated acute myocardial infarction. They were considered so important in causing this acute event that the term "coronary thrombosis" was used for years to describe the event that more physicians now call "acute myocardial infarction". To evaluate the significance of coronary thrombus in acute myocardial infarction, Brosius and Roberts (1981) examined in detail the coronary arteries containing thrombi in 54 autopsied patients with transmural acute myocardial infarction. Of 235 patients with fatal acute myocardial infarction studied, 99 had histological sections available from each 5-mm segment of each of the four major coronary arteries. Movat-stained histological sections, approximately 55 per patient, were reviewed, and a thrombus was found in one of the four major coronary arteries in 54 patients (55%). In the coronary artery containing the thrombus, the maximal degree of cross-sectional area narrowed by atherosclerotic plaque was determined at the site of the thrombus, in the 2-cm portion of artery proximal to the thrombus, and in the 2-cm segment of coronary artery distal to the distal site of attachment of the thrombus. The length of the thrombus was determined by the number of 5-mm long segments of coronary artery that contained thrombus.

A coronary arterial thrombus was defined as a collection of fibrin (with or without engulfed erythrocytes) or platelets or both within the residual lumen with attachment of the fibrin/platelets to the luminal surface of the artery. Among the 54 patients with fatal acute myocardial infarction, the luminal surface was always the surface of an underlying atherosclerotic plaque. The thrombus was always attached to the intimal surface in its distal portion, but in a few patients it was not attached in its most proximal portion. The thrombus was considered occlusive when it occupied totally the residual lumen of the artery, i.e. the portion not occupied by atherosclerotic plaque. The thrombus was considered non-occlusive when it filled the residual lumen incompletely. The area of the occlusive thrombus was the difference between the area of the original lumen and the area of the atherosclerotic plaque. In the case of non-occlusive thrombus, the residual lumen was the difference between the artery's original lumen and the sum of the area occupied by atherosclerotic plaque plus the area occupied by the non-occlusive thrombus. In the sections of coronary artery proximal and distal to the thrombus, the residual lumen was the difference between the artery's original lumen and the area occupied by atherosclerotic plaque. The percentage of cross-sectional area narrowed by atherosclerotic plaque and by thrombi (in the case of non-occlusive thrombi) was calculated. The area of each artery enclosed by the internal elastic membrane (original lumen), the area of the atherosclerotic plaque, and that of the non-occlusive thrombus provided by videoplanimetry were converted into actual areas.

All patients had had acute myocardial infarction that involved the entire inner half of the left ventricular wall and a portion or all of the outer half of the left ventricular wall. Twenty patients also had one or more transmural scars. The 54 patients ranged in age from 38 to 92 years (mean 62 years); 41 were men and 13 were women.

The thrombi were occlusive in 47 patients (87%) and non-occlusive in seven (13%). The coronary arterial systems containing thrombi were the left anterior descending in 22 (41%), the right in 19 (35%), and the left circumflex in 13 (24%). The coronary thrombi ranged from 0.5 cm to 10 cm in length (mean 1.6 cm). In the right coronary artery, the length of the thrombus increased directly as the intervals increased between onset of acute myocardial infarction and death. The distance from the origin of a coronary artery (aorta for the right and left main for the left anterior descending and left circumflex arteries) to the most proximal portion of a coronary thrombus ranged from zero to 15.5 cm (mean 2.9 cm). The mean distance of an occlusive thrombus from the aorta in the right coronary artery, however, was significantly greater than the mean distance of either an occlusive or non-occlusive thrombus from the left main coronary artery in either the left anterior descending or left circumflex coronary arteries.

The maximal luminal narrowing by atherosclerotic plaques alone at the site of the thrombus varied from 33% to 98% (mean 81%); the maximal coronary luminal narrowing by thrombus alone varied from 2% to 67% (mean 19%) in the 47 patients with occlusive thrombi and from 2% to 24% (mean 7%) in the seven patients with non-occlusive thrombi. The maximal luminal narrowing by atherosclerotic plaques in the 2 cm of coronary artery proximal to the thrombus ranged from 26% to 98% (mean 75%), and in the 2 cm distal to the distal site of attachment of the thrombus from 43% to 98% (mean 76%). No significant differences were noted in the amount of maximal luminal narrowing at, proximal to, or distal to the thrombus between the 47 patients with occlusive thrombi and the seven with non-occlusive thrombi. Likewise, no significant differences were observed in the amount of maximal luminal narrowing at, proximal to, or distal to the coronary thrombus in the three major coronary arteries.

The absolute area occupied by atherosclerotic plaques, the area occupied by thrombus, and the area of the residual coronary lumina for the sites of maximal narrowing at, proximal to, and distal to coronary artery thrombi were as follows: at the site of maximal luminal narrowing of the coronary artery at the site of the thrombus, the area occupied by atherosclerotic plaque ranged from $1.5 \, mm^2$ to $15.0 \, mm^2$ (mean $7.2 \, mm^2$); the area occupied by occlusive thrombus from $0.1 \, mm^2$ to $6.0 \, mm^2$ (mean $1.5 \, mm^2$); and that occupied by non-occlusive thrombus from $0.1 \, mm^2$ to $1.5 \, mm^2$ (mean $0.5 \, mm^2$). The area of the original coronary lumen (that enclosed by internal elastic membrane) at the site of the thrombus ranged from $3 \, mm^2$ to $21 \, mm^2$ (mean $8.7 \, mm^2$); proximal to the thrombus from $2.8 \, mm^2$ to $13.6 \, mm^2$ (mean $8.7 \, mm^2$); and distal to the thrombus from $0.8 \, mm^2$ to $13.5 \, mm^2$ (mean $5.1 \, mm^2$). The maximal area occupied by atherosclerotic plaque proximal to the thrombus ranged from $2.2 \, mm^2$ to $12.1 \, mm^2$ (mean $6.5 \, mm^2$); and distal to the thrombus from $0.4 \, mm^2$ to $11.8 \, mm^2$

(mean $4.0 \, mm^2$). No significant differences in mean areas between the 47 patients with occlusive and the seven with non-occulsive coronary thrombi were observed in the original size of the coronary artery or in the amount of lumen obliterated by atherosclerotic plaque proximal to, at, or distal to the coronary thrombus.

In 52 (96%) of the 54 coronary arteries with thrombi, the lumina of the arteries at, proximal to, or distal to the thrombi were already narrowed 76%–100% in cross-sectional area by atherosclerotic plaques: in 27 of 42 arteries (64%) examined proximal to the thrombus (in 12 arteries the thrombi began at, or nearly at, the origin of the coronary artery from the aorta or left main coronary artery), in 35 (66%) of 53 arteries examined distal to the distal site of attachment of the thrombus, and in 41 (76%) of the 54 coronary arteries (or patients) at the site of attachment of the thrombus. Furthermore, in 26 (48%) of the 54 coronary arteries, the lumina of the arteries at, proximal to, or distal to the thrombi were narrowed 91%–98% in cross-sectional area by atherosclerotic plaques. In 16 (30%) of 54 coronary arteries with thrombi, the site of most severe narrowing by atherosclerotic plaque in the portion of artery examined was within the 2 cm proximal to the thrombi: in 25 of 54 (46%) it was at the site of the thrombus and in 13 of 54 (24%) it was in the 2 cm segment distal to the thrombus. The site of most severe narrowing in the coronary arteries was not significantly different between occlusive and non-occlusive thrombi.

At the site of attachment of thrombi, the underlying atherosclerotic plaques contained extravasated erythrocytes in 21 (39%) of the 54 patients. In none of the 21, however, did the hemorrhage into the pultaceous debris of the plaque appear to compromise the lumen.

The Brosius–Roberts study summarized above raises questions regarding the importance of coronary thrombi in patients with fatal transmural acute myocardial infarction. The major finding at autopsy is that, among patients with fatal acute myocardial infarction, thrombi are found in major coronary arteries that already are severely narrowed by old atherosclerotic plaques at, immediately proximal to, and/or immediately distal to the site of thrombosis. The lumen of the coronary artery containing the thrombus was already narrowed an average of 79% (range 26%–98%) in cross-sectional area by atherosclerotic plaque alone at and within 2 cm proximal to and distal to the thrombus; that is, an "average" coronary artery with a thrombus was severely narrowed (75% in cross-sectional area) at three sites (at, proximal to, and distal to the thrombus). The average coronary arterial narrowing at the site of thrombus, however, actually underestimates the true severity of the narrowing in the vicinity of the thrombus. The site of most severe narrowing was within the 2 cm proximal to the thrombus in 16 of 54 (30%) coronary arteries, at the site of thrombus in 25 (46%), and within the 2-cm segment distal to the thrombus in 13 (24%). At the site of most severe narrowing, 96% of the coronary arteries were narrowed 76%–98% in cross-sectional area by atherosclerotic plaque, and half were narrowed 91%–98%. In contrast, the percentage of coronary lumen narrowed by thrombus alone averaged 19% of the original cross-sectional area of the artery (range 2%–67%) in the 47 patients with occlusive thrombi, and 7% (range 2%–24%) in the seven patients with non-occlusive thrombi. Thus, if thrombus were the only luminal material, the amount of thrombus within the coronary artery, with a few exceptions, probably would not by itself diminish or slow blood flow. Thus, the frequent need for bypass surgery or angioplasty after successful thrombolysis is readily understandable. A corollary is that among necropsy patients with fatal acute myocardial infarction, the coronary thrombus, when present, is always superimposed on an atherosclerotic plaque. The exception is coronary embolism, when clot may be present without underlying atherosclerotic plaque (Roberts 1978).

The length of coronary thrombi in necropsy patients with fatal acute myocardial infarction is usually short. Among the 54 patients studied by Brosius and Roberts, the average coronary thrombus was 1.6 cm long (range 0.5–10 cm); the occlusive thrombi were longer than the non-occlusive thrombi (1.8 cm vs. 0.7 cm). Also occlusive thrombi in the right coronary arteries tended to be longer than those in the left anterior and circumflex coronary arteries (2.4 cm vs. 1.4 cm and 1.1 cm). Because the right and left anterior descending coronary arteries in adults are more than 10 cm long and the left circumflex is usually about 6 cm long, the actual length of a coronary artery occupied by thrombus is small, and in no patient was the entire length of a coronary artery occupied by a thrombus. In the right coronary artery, there was a weak, but significant, positive correlation between the length of an occlusive thrombus and the duration of survival of a patient between the time of acute myocardial infarction and death. This relation suggests that thrombi may lengthen or "grow" with time.

Cardiac Morphological Findings in Acute Myocardial Infarction Treated with Thrombolytic Therapy

Until thrombolytic and revascularization therapy of acute myocardial infarction was introduced, virtually all acute myocardial infarcts observed at necropsy were non-hemorrhagic. Treatment with thrombolytic agents has been shown to restore the patency of occluded coronary arteries, but this has been associated with an apparently marked, although heretofore undetermined, increase in the frequency of hemorrhagic infarcts. It has been suggested that myocardial hemorrhage after coronary reperfusion is confined to zones of the myocardium that were already necrotic and that the hemorrhage is probably a consequence of severe microvascular injury and not its cause. Gertz and associates (1990a) studied at necropsy the hearts of 52 patients who had received recombinant tissue plasminogen activator (rt-PA) during acute myocardial infarction and compared clinical and cardiac morphological findings in patients with hemorrhagic infarcts to those with non-hemorrhagic infarcts. The acute infarcts were hemorrhagic by gross inspection (with histological confirmation) in 23, non-hemorrhagic in 20, not visible grossly in 2 and, in 7, there was no acute necrosis by either gross or histological examination of multiple sections of the myocardium. In 4 of these 7 patients without acute infarcts, the interval from chest pain to death was <10 hours, which is often too early to detect the presence of necrosis by histological examination.

No significant differences were found between patients with hemorrhagic and non-hemorrhagic infarcts with respect to mean age, heart weight, interval from chest pain to rt-PA infusion, interval from chest pain to peak creatinine kinase, interval from chest pain to death, location of the myocardial necrosis, frequency of left ventricular dilatation, frequency of myocardial rupture (left ventricular free wall or ventricular septum) or frequency of cardiogenic shock, fatal arrhythmias or fatal bleeding. Furthermore, the frequencies of thrombi and plaque rupture in coronary arteries and the sizes of the infarct were similar in patients with hemorrhagic and non-hemorrhagic infarcts. Thus, although the frequency of hemorrhagic infarction increases after thrombolytic therapy, the hemorrhage does not appear to extend the infarct or to increase the frequency of complications of infarct.

In a separate study, Gertz and associates (1990b) compared cardiac findings at necropsy in 23 patients who had received thrombolytic therapy during acute myocardial infarction to those in 38 patients with acute myocardial infarction who had not received thrombolytic therapy. Although each group of patients had similar baseline characteristics, the patients receiving thrombolytic therapy (rt-PA) had a greater frequency of platelet-rich (fibrin-poor) thrombi in the infarct-related coronary arteries, more non-occlusive than occlusive thrombi, and a lower frequency of myocardial rupture.

Composition of Atherosclerotic Plaques in Fatal Coronary Artery Disease

Kragel and associates (1989, 1990) studied at necropsy atherosclerotic plaque composition in the four major (right, left main, left anterior descending, and left circumflex) coronary arteries in 15 patients who died of consequences of acute myocardial infarction, in 12 patients with sudden coronary death without associated myocardial infarction, and in 10 patients with isolated unstable angina pectoris with pain at rest. The coronary arteries were sectioned at 5-mm intervals, and a Movat-stained section of each segment of artery was prepared and analyzed using a computerized morphometry system. Among the three subsets of coronary patients, there were no differences in plaque composition among any of the four major epicardial coronary arteries. Within all three groups the major component of plaque was a combination of dense acellular and cellular fibrous tissue with much smaller portions of plaque being composed of pultaceous debris, calcium, foam cells with and without inflammatory cells, and foam cells alone (Fig. 1.3). Within all three groups, plaque morphology varied as a function of cross-sectional area narrowing of the segments. In all three groups, the amount of dense, relatively acellular fibrous tissue, calcified tissue, and pultaceous debris (amorphous debris containing cholesterol clefts, presumably rich in extracellular lipid) increased in a linear fashion with increasing degrees of cross-sectional area narrowing of the segments and the amount of cellular fibrous tissue decreased linearly. The percentage of plaque consisting of pultaceous debris was highest in the subgroup with acute myocardial infarction. Mutiluminal channels were most frequent in the subgroup with unstable angina pectoris. The studies by Kragel and associates (1989, 1990) are the first to analyze quantitatively the composition of coronary arterial plaques in the various subsets of coronary patients.

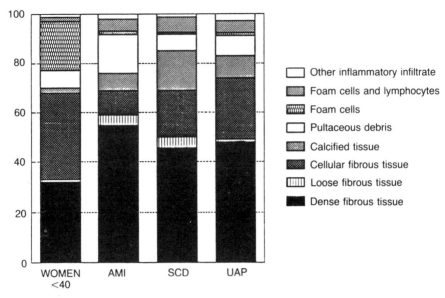

Fig. 1.3. Bar graph showing atherosclerotic plaque composition (mean percentage) in histological sections of 5-mm sections of the four major epicardial coronary arteries narrowed greater than 75% in cross-sectional area. The patients included 8 women aged 31–39 years (mean 34) with fatal coronary artery disease and 35 patients over 40 years of age (mean 59 years) with fatal acute myocardial infarction (AMI), sudden coronary death (SCD) and unstable angina pectoris (UAP). The dominant component of the atherosclerotic plaques in all groups was fibrous tissue. (Reproduced with permission from Dollar et al. (1991) Composition of atherosclerotic plaques in coronary arteries in women <40 years of age with fatal coronary artery disease and implications for plaque reversibility. *Am. J. Cardiol.* 67.)

Effects of Percutaneous Transluminal Coronary Angioplasty on Atherosclerotic Plaques and Relation of Plaque Composition and Arterial Size to Outcome

To delineate their relation to outcome of percutaneous transluminal coronary angioplasty (PTCA), the atherosclerotic plaque composition and coronary artery size in 82 5-mm long segments at 28 PTCA sites were determined by Potkin and Roberts (1988) in 26 patients having PTCA. The 26 patients were subdivided into three groups according to the degree of angiographic patency at the end of the PTCA procedure and to the duration of survival after PTCA ($\leqslant$30 or >30 days): *early success* (13 patients, 16 PTCA sites and 49 5-mm segments); *early failure* (4 patients, 4 PTCA sites and 16 5-mm segments) and *late success* (9 patients, 8 PTCA sites and 17 5-mm segments). The mean percentage of plaque comprised of fibrous tissue among the three groups was 80% $\pm$ 18%, 71% $\pm$ 23% and 82% $\pm$ 16%; the mean percentage of plaque comprised of lipid was 17% $\pm$ 16%, 21% $\pm$ 24% and 16% $\times$ 15%; and of calcium it was 3% $\pm$ 4%, 8% $\pm$ 10% and 2% $\pm$ 3%. The mean coronary arterial internal diameter was 3.3 $\pm$

0.6 mm, 3.9 ± 1.2 mm and 3.2 ± 0.7 mm. Plaque tear was present in one or more histological sections in 25 of the 26 patients and the one patient without it had the longest interval (nearly three years) between PTCA and death. Plaque tear extending from intima into media with dissection was observed only in the early and late success groups. Hemorrhage into plaque was present in 16 (80%) of 20 PTCA sites in the two early groups and in three (37%) of eight sites in the late group. Occlusive thrombus (5 of 16, 1 of 4 and 1 of 8) and plaque debris (7 of 16, 1 of 4 and 2 of 8) in residual lumina were insignificantly different among the three groups and their 82 5-mm segments. Plaques that had >25% lipid content had an increased frequency of hemorrhage into plaque, occlusive thrombus and plaque debris in residual lumens. These findings suggest that coronary arterial size and plaque composition are strong determinants of PTCA outcome. The ideal coronary arterial atherosclerotic narrowing for both technically and clinically successful PTCA appears to be a small (<3.3 mm in internal diameter) artery in which the plaque contains relatively little calcium and lipid. This study by Potkin and Roberts (1988) was the first to quantify the composition of atherosclerotic plaque at the site of successful and unsuccessful PTCA, and also the first to measure the coronary arterial internal diameter at sites of PTCA.

Morphological Findings in Saphenous Veins Used as Coronary Arterial Bypass Conduits

Saphenous veins, when used as aorto-coronary conduits, undergo changes in their intimal, medial, and adventitial layers. The predominant late intimal change is a proliferation of fibrous tissue, a finding observed within two months after coronary artery bypass grafting. Other late changes in saphenous vein grafts include deposits of lipid, thrombus, and rarely aneurysm formation. Most published studies describing changes in saphenous veins used as bypass conduits have involved few necropsy patients, involved only operatively excised specimens, or involved cases with relatively short intervals from coronary bypass to death or reoperation. Kalan and Roberts (1990) studied at necropsy the hearts and grafts of 53 patients who lived longer than one year after coronary bypass. They examined 123 saphenous vein grafts and 1865 5-mm segments of the grafts in the 53 patients, some of whom died of consequences of myocardial ischemia and some of whom died of non-cardiac conditions.

The 53 patients died from 13 to 53 months (mean 58) after a single aorto-coronary bypass operation. Of the 53 patients, 32 (60%) died of a cardiac cause and, of their 72 saphenous vein aorto-coronary conduits, 36 (49%) were narrowed at some point more than 75% in cross-sectional area by atherosclerotic plaque; the remaining 21 patients (40%) died of a non-cardiac cause, and, of their 50 saphenous vein conduits, 10 (20%) were narrowed at some point more than 75% in cross-sectional area by plaque. Thus, the non-cardiac mode of death in a large percentage of the patients suggests that the bypass operation prolonged life to a degree sufficient for another condition to develop. The 123 saphenous vein conduits were divided into 5-mm segments, and a histological section was prepared from each. Of the 1104 5-mm segments in the 32 patients dying as a consequence of myocardial ischemia, 291 (26%) were narrowed more than 75% in cross-sectional area by plaque; in contrast, of the 761 5-mm segments of veins in the 21 patients with a non-cardiac mode of death, 86 (11%) were narrowed more than 75% by plaque. Of the total 1865 5-mm segments of vein, only 395 (21%) were narrowed 25% or less in cross-sectional area by plaque. Thus, in patients dying late after coronary bypass the atherosclerotic process continues in all segments of the saphenous veins used as aorto-coronary conduits. Therapy after the operation must be directed toward prevention of progression of the atherosclerosis in the "new" coronary "arteries".

In the study by Kalan and Roberts (1990), the amount of luminal narrowing in the saphenous veins used as aorto-coronary conduits was significantly greater in those patients who died of a cardiac cause compared to those who died of a non-cardiac cause. Additionally, the percentage of vein conduits and the percentage of 5-mm segments of the vein conduits totally occluded or nearly so (>95% in cross-sectional area) were significantly greater in the patients dying of a cardiac cause compared with those dying of a non-cardiac cause (22 (30%) of 73 veins vs. 7 (14%) of 50 veins, and 152 (14%) of 1104 segments vs. 57 (27%) of 213 segments).

Surprisingly, the interval from coronary bypass to death did not correlate with either the percentage of vein conduits or the percentage of 5-mm segments of vein conduit narrowed more than 75% in cross-sectional area by plaque. The percentage of venous conduits narrowed severely (>75%) was similar in the 18 conduits (7 patients) in place from 13 to 24 months and in the 9 conduits (5 patients) in place for longer than ten years. Moreover, the percentage of 5-mm segments of saphenous vein conduit severely narrowed was similar in the 35 patients surviving up to five years compared to the 18 patients surviving more than five years (268 (19%) of 1387 segments vs. 113 (24%) of 478 segments).

Why some saphenous vein conduits became severely narrowed or occluded or nearly so and others did not may be related more to the status of the native coronary artery containing the graft than to the graft itself. Of the 123 native coronary arteries containing a saphenous vein conduit, 49 (40%) of the arteries distal to the anastomotic site were narrowed more than 75% in cross-sectional area by plaque, and the anastomosed saphenous vein was severely narrowed in 33 (67%) of them. In contrast, of the 74 native coronary arteries narrowed less than 75% distal to the anastomotic site, the attached saphenous vein was severely narrowed in only 14 (19%). Thus the amount of narrowing in the native coronary artery distal to the anastomotic site plays a major determining role in the fate of the attached saphenous vein.

The composition of the plaques in the saphenous venous conduits is similar to that in the native coronary arteries. Fibrous tissue or fibromuscular tissue was the dominant component of the plaques in the saphenous vein conduits just as it is the dominant component of plaques in the native coronary arteries in patients with fatal coronary artery disease without coronary bypass. Lipid was present in plaques in saphenous veins in much smaller amounts than was fibrous tissue. Intracellular lipid was found in a saphenous vein as early as 14 months after coronary bypass, and it did not increase in either frequency or amount as the interval from bypass to death increased. Extracellular lipid was first seen at 26 months after bypass, and it did not appear to increase thereafter as the interval from bypass increased. Hemorrhage into plaque, which occurred almost entirely into extracellular lipid deposits (containing cholesterol clefts within pultaceous debris), was first seen at 32 months after bypass. Intraluminal thrombus was found in saphenous veins in 14 patients (26%) and was first observed at 32 months. Thrombus was always superimposed on underlying lipid plaque. Calcific deposits were found in saphenous vein conduits in 11 patients (21%); they were first noted at 34 months after bypass, and they did increase in frequency with time.

The frequency of the various modes of death among the patients dying late after coronary bypass is a bit different from that of patients with symptomatic myocardial ischemia without coronary bypass. Of the 53 coronary bypass patients studied, only 32 (60%) died of a cardiac cause and therefore 21 (40%) died of a non-cardiac cause. Among patients with symptomatic myocardial ischemia who do not have coronary bypass, approximately 95% die of a cardiac cause, and therefore only about 5% die of a non-cardiac cause. The fact that 40% of the bypass cases studied died of a non-cardiac cause supports the view that the bypass operation in many patients prolongs life long enough in many for various fatal non-cardiac conditions to develop. Of the 53 bypass patients studied by Kalan and Roberts (1990) 10 (19%) died of cancer, a percentage far higher than patients with symptomatic myocardial ischemia not having coronary surgery.

The Kalan–Roberts study re-emphasizes that coronary bypass is useful but that it does not deter progression of the underlying atherosclerotic process. In a slight way the bypass operation might even cause acceleration of the atherosclerotic process because in about 25% of persons having coronary bypass the serum total cholesterol increases and the body weight increases substantially during the first year after operation. Because lowering the serum (or plasma) total cholesterol level (and specifically the low-density lipoprotein cholesterol) causes some portion of atherosclerotic plaques to regress and the chances of a fatal or non-fatal subsequent atherosclerotic event to decrease, a strong case can be advanced for combined simultaneous initiation of both low-fat, low-cholesterol diet therapy and lipid-lowering drug therapy as soon as is reasonably feasible after a coronary bypass operation (Roberts 1990).

References

Arnett EN, Isner JM, Redwood DR, Kent KM, Baker WP, Ackerstein M, Roberts WC (1979) Coronary artery narrowing in coronary heart disease: comparison of cineangiographic and necropsy findings. Ann Intern Med 91:350–356

Brosius FC III, Roberts WC (1981) Significance of coronary arterial thrombus in transmural acute myocardial infarction: a study of 54 necropsy patients. Circulation 63:810

Bulkley BM, Roberts WC (1976) Atherosclerotic narrowing of the left main coronary artery: a necropsy analysis of 152 patients with fatal coronary heart disease and varying degrees of left main narrowing. Circulation 53:823–828

Cabin HS, Roberts WC (1980) True left ventricular aneurysm and healed myocardial infarction. Clinical and necropsy observations including quantification of degree of coronary arterial narrowing. AM J Cardiol 46:754

Gertz SD, Kalan JM, Kragel AH, Roberts WC, Braunwald E, The TIMI Investigators (1990a) Cardiac morphologic findings in patients with acute myocardial infarction treated with recombinant tissue plasminogen activator. Am J Cardiol 65:953–961

Gertz SD, Kragel AH, Kalan JM, Braunwald E, Roberts WC, The TIMI Investigators (1990b) Comparison of coronary and myocardial morphologic findings in patients with and without thrombolytic therapy during fatal first acute myocardial infarction. Am J Cardiol 66:904–909

Herrick JB (1919) Thrombosis of the coronary arteries. JAMA 72:387

Isner JM, Kishel J, Kent KM, Ronan JA Jr, Ross AM, Roberts WC (1981) Accuracy of angiographic determination of left main coronary arterial narrowing. Angiographic–histologic correlative analysis in 28 patients. Circulation 63:1056–1064

Kalan JM, Roberts WC (1990) Morphologic findings in saphenous veins used as coronary arterial bypass conduits for longer than 1 year: necropsy analysis of 53 patients, 123 saphenous veins, and 1865 5-mm segments of veins. Am Heart J 119:1164–1184

Kragel AH, Reddy SG, Wittes JT, Roberts WC (1989) Morphometric analysis of the composition of atherosclerotic plaques in the four major epicardial coronary arteries in acute myocardial infarction and in sudden coronary death. Circulation 80:1747–1756

Kragel AH, Reddy SG, Wittes JT, Roberts WC (1990) Morphometric analysis of the composition of coronary arterial plaques in isolated unstable angina pectoris with pain at rest. Am J Cardiol 66:562–567

Kragel AH, Gertz SD, Roberts WC (1991) Morphologic comparison of frequency and types of acute lesions in the major epicardial coronary arteries in unstable angina pectoris, sudden coronary death, and acute myocardial infarction. J Am Coll Cardiol 18

Potkin BN, Roberts WC (1988) Location of an acute myocardial infarct in patients with a healed myocardial infarct: analysis of 129 patients studied at necropsy. Am J Cardiol 62:1017–1023

Roberts WC (1978) Coronary embolism. A review of causes, consequences, and diagnostic considerations. Cardiovasc Med 3:699

Roberts WC (1989) Qualitative and quantitative comparison of amounts of narrowing by atherosclerotic plaques in the major epicardial coronary arteries at necropsy in sudden coronary death, transmural acute myocardial infarction, transmural healed myocardial infarction and unstable angina pectoris. Am J Cardiol 64:324–328

Roberts WC (1990) Lipid-lowering therapy after an atherosclerotic event. Am J Cardiol 65:16F–18F

Roberts WC, Buja LM (1972) The frequency and significance of coronary arterial thrombi and other observations in fatal acute myocardial infarction: a study of 107 necropsy patients. Am J Med 52:425

Roberts WC, Jones AA (1979) Quantitation of coronary arterial narrowing at necropsy in sudden coronary death: analysis of 31 patients and comparison with 25 control subjects. Am J Cardiol 44:39

Roberts WC, Jones AA (1980) Quantification of coronary arterial narrowing at necropsy in acute transmural myocardial infarction: analysis and comparison of findings in 27 patients and 22 controls. Circulation 61:786

Roberts WC, Virmani R (1979) Quantification of coronary arterial narrowing in clinically isolated unstable angina pectoris: an analysis of 22 necropsy patients. Am J Med 67:792

Virmani R, Roberts WC (1980) Quantification of coronary arterial narrowing and of left ventricular myocardial scarring in healed myocardial infarction with chronic eventually fatal congestive cardiac failure. Am J Med 68:831

Virmani R, Roberts WC (1981) Non-fatal healed transmural myocardial infarction and fatal non-cardiac disease: qualification and quantification of coronary arterial narrowing and of left ventricular scarring in 18 necropsy patients. Br Heart J 45:434

Waller BF, Roberts WC (1980) Amount of narrowing by atherosclerotic plaque in 44 nonbypassed and 52 bypassed major epicardial coronary arteries in 32 necropsy patients who died within 1 month of aortocoronary bypass grafting. Am J Cardiol 46:956

Wartman WB, Hellerstein HK (1948) The incidence of heart disease in 2000 consecutive autopsies. Ann Intern Med 28:41

Chapter 2

The L-Arginine:Nitric Oxide Pathway in Physiology and Pathology

S. Moncada and E. A. Higgs

Introduction

The demonstration that vascular endothelial cells form nitric oxide (NO) from the amino acid L-arginine has led to the discovery that this molecule not only explains the biological properties of the so-called endothelium-derived relaxing factor (EDRF) but is also the stimulator of the soluble guanylate cyclase in a number of tissues such as the platelet and the brain. Furthermore, NO is a cytotoxic factor released by activated murine macrophages and other cells. NO is synthesized by an enzyme, NO synthase, of which two types have been identified, one constitutive and the other induced by cytokines.

The Constitutive Nitric Oxide Synthase

The Vasculature

EDRF (Furchgott 1984) is a labile humoral substance which accounts for the vascular relaxation induced by acetylcholine (ACh) and other endothelium-dependent vasodilators. It also inhibits platelet aggregation (Azuma et al. 1986; Radomski et al. 1987a) and adhesion (Radomski et al. 1987b), all of which actions are mediated via stimulation of the soluble guanylate cyclase. The chemical nature of EDRF has now been identified as NO since the pharmacological properties of EDRF and NO are identical (for review see Moncada et al. 1988a) and NO is released from vascular endothelial cells in culture (Palmer et al. 1987; Kelm et al. 1988) and from vascular preparations

(Ignarro et al. 1987; Khan and Furchgott 1987; Amezcua et al. 1988; Kelm and Schrader 1988; Chen et al. 1989) in amounts sufficient to explain the biological actions of EDRF.

Porcine vascular endothelial cells in culture synthesize NO from the terminal guanidino nitrogen atom(s) of L-arginine (Palmer et al. 1988a; Schmidt et al. 1988). This reaction is specific, since a number of analogs of L-arginine, including its D-enantiomer, are not substrates. In addition, one analog, N^G-monomethyl-L-arginine (L-NMMA), inhibits this synthesis in a dose dependent and enantiomerically specific manner (Palmer et al. 1988b). The co-product of this reaction is L-citrulline, for its synthesis from L-arginine by vascular endothelial cell homogenates is also inhibited by L-NMMA (Palmer and Moncada 1989).

The enzyme that synthesizes NO from L-arginine has been studied in endothelial homogenates and characterized (Palmer and Moncada 1989). This enzyme, which has now been called the NO synthase (Moncada and Palmer 1990), is NADPH and Ca^{2+}/calmodulin dependent (Mayer et al. 1989; Mulsch et al. 1989).

In rabbit aortic rings the NO synthase inhibitor L-NMMA induces a small but significant endothelium-dependent contraction and inhibits the relaxation and the release of NO induced by acetylcholine (ACh). L-Arginine, which on its own only induces a small endothelium-dependent relaxation, antagonizes all the actions of L-NMMA (Rees et al. 1989a). Similar results have been obtained in guinea-pig pulmonary artery rings (Sakuma et al. 1988) and in the coronary circulation of the rabbit heart in vitro (Amezcua et al. 1989).

In the anesthetized rabbit L-NMMA induces a dose-dependent, long-lasting increase in mean arterial blood pressure and inhibits the hypotensive action of ACh (glyceryl trinitrate (GTN), Rees et al. 1989b). These effects of L-NMMA are reversible by L-arginine. L-NMMA has subsequently been shown to cause a rise in blood pressure in anesthetized guinea-pigs (Aisaka et al. 1989) and rats (Whittle et al. 1989; Gardiner et al. 1990a; Tolins et al. 1990; Rees et al. 1990a). In one of these studies (Tolins et al. 1990), the vasodilatation induced by ACh was accompanied by an increase in the urinary excretion of cyclic GMP, both of which were prevented by L-NMMA. The effects of L-NMMA on blood pressure were also associated with a decrease in glomerular filtration rate.

The increase in blood pressure induced by L-NMMA was also associated with a decrease in vascular conductance in the renal, mesenteric, carotid and hindquarters vascular beds of conscious, chronically instrumented rats (Gardiner et al. 1990b). Furthermore, these effects were sustained if the infusion of L-NMMA was continued for up to 6 hours (Gardiner et al. 1990c), indicating not only the critical role of NO in maintaining vascular patency in all these beds, but also the fact that regulatory systems in the vasculature are unable to reaccommodate the flow towards pretreatment levels. In awake, chronically instrumented dogs, L-NMMA induced a dose-related, L-arginine-reversible constriction of the coronary circulation together with a reduction in resting phasic coronary flow (Chu et al. 1991). Preliminary evidence suggests that the coronary vasodilatation which follows vagal stimulation is also NO dependent (Broten et al. 1991).

Studies in which L-NMMA has been infused in man into the brachial artery or the dorsal veins of the hand demonstrated that the vasodilatation induced by ACh or bradykinin, but not that induced by GTN, could be attenuated by this compound (Vallance et al. 1989a, 1989b). Furthermore, while in the brachial artery L-NMMA induced direct vasoconstriction, it had no such direct effect on the hand veins. This suggests that in the arterial side of the circulation, but not the venous side, there is a continuous release of NO which maintains a dilator tone. That the arterial side of the circulation releases, in general, more NO than the venous side is also suggested by the fact that ACh-induced dilatation in veins was rapidly transformed into constriction as the dose increased. Interestingly, the dilatation was attenuated by L-NMMA while the constriction was enhanced, suggesting that NO mediates, at least in part, the dilatation and functionally antagonizes vasoconstrictor responses (Vallance et al. 1989b).

The L-arginine analogs N^G-nitro-L-arginine (L-NNA), its methyl ester L-NAME and N-iminoethyl-L-ornithine (L-NIO) have all recently been described as inhibitors of NO generation in vascular tissue (Rees et al. 1990a, 1990b; Moore et al. 1989; Ishii et al. 1990; Mulsch and Busse, 1990; Fukuto et al. 1990; Gardiner et al. 1990d). Interestingly, L-NIO was approximately five times more potent than the other analogs, suggesting that there may be differences in uptake, distribution or metabolism of these compounds (Rees et al. 1990a). Some of these compounds are orally active, for L-NMMA and L-NAME, given by this route to Brattleboro rats, induced a sustained increase in blood pressure (Gardiner et al. 1990d).

All these results clearly indicate that there is, in the vasculature, a continuous utilization of L-arginine for the generation of NO which plays a role in the maintenance of blood pressure. The release of NO can be enhanced further by endothelium-dependent vasodilators. Furthermore, the marked rise in blood pressure obtained after inhibition of NO synthesis confirms the proposal that NO is the endogenous nitrovasodilator (for review see Moncada et al. 1988b) and suggests that a reduction in the synthesis of NO may contribute to the pathogenesis of hypertension.

The Platelet

We have recently shown that platelets also generate NO and that the L-arginine:NO pathway acts as a negative feedback mechanism to regulate platelet aggregation (Radomski et al. 1990a). Aggregation induced by collagen was accompanied by an increase in intra-platelet levels of cyclic GMP but not cyclic AMP. L-NMMA inhibited this increase in cyclic GMP and enhanced aggregation. Furtermore, L-arginine, which had no effect on basal levels of cyclic GMP, enhanced the increase in cyclic GMP induced by collagen and inhibited aggregation.

An increase in cyclic GMP in platelet cytosol was observed not only with sodium nitroprusside (SNP, a generator of NO) but also with L-arginine. The effect of L-arginine was enantiomer specific, inhibited by L-NMMA and dependent on the presence of NADPH. In addition, measurements in the platelet cytosol demonstrated an L-arginine- and NADPH-dependent formation of NO which was inhibited by L-NMMA, providing conclusive evidence for the existence of the L-arginine:NO pathway in the platelets. The formation of NO from L-arginine in platelet cytosol was dependent on the free Ca^{2+} concentration, showing that the NO synthase in platelets, as in the vascular endothelium, is Ca^{2+} dependent.

Thus, platelet aggregation in vivo is likely to be regulated by intra-platelet NO, as well as by NO and prostacyclin released from vascular endothelium. The combined action of these two mediators could result in a synergistic suppression of elevation of intracellular Ca^{2+} and subsequent powerful inhibition of platelet aggregation (Radomski et al. 1987c).

The Nervous System

In 1977 NO was shown to stimulate the soluble guanylate cyclase in homogenates of mouse cerebral cortex (Miki et al. 1977). In the same year, Deguchi showed that the soluble fraction of rat forebrain contained a low-molecular-weight substance which activated soluble guanylate cyclase and whose action was inhibited by hemoglobin (Deguchi 1977). Similar findings were reported in the rat cerebellum (Yoshikawa and Kuriyama 1980). In 1982 the endogenous activator of the soluble guanylate cyclase in the brain was identified as L-arginine (Deguchi and Yoshioka 1982).

These observations, together with the discovery of the L-arginine:NO pathway in the vascular endothelium, led us to investigate the existence of this pathway in the central nervous sytem to determine whether it accounted for L-arginine-dependent stimulation of the soluble guanylate cyclase. Addition of L-arginine to rat synaptosomal cytosol in the presence of NADPH resulted in the formation of NO and citrulline accompanied by stimulation of soluble guanylate cyclase (Knowles et al. 1989). Both these processes were inhibited by hemoglobin and by L-NMMA, but not by L-canavanine. These data showed that the rat brain possesses the NO synthase. This enzyme was dependent on the free Ca^{2+} concentration (Knowles et al. 1989) so that it was essentially inactive at the resting free Ca^{2+} concentration in synaptosomes (approx. 80 nM; Ashley et al. 1984), whereas it was fully active at Ca^{2+} concentrations around 400 nM. As in the vascular endothelium and the platelet, therefore, increases in intracellular Ca^{2+} may constitute the physiological mechanism for stimulating the synthesis of NO.

Garthwaite et al. (1988) had reported that stimulation of rat cerebellar cells with N-methyl-D-aspartate (NMDA) induced an elevation of cyclic GMP levels which was associated with the release of an EDRF-like material. Furthermore, the cells that released this EDRF-like material in response to excitatory stimulation were not the target cells in which cyclic GMP levels were elevated. The cyclic GMP response to NMDA stimulation was later shown to be enhanced by L-arginine and inhibited by L-NMMA in a manner that was reversed by additional L-arginine (Garthwaite et al. 1989a; Bredt and Snyder 1989), showing that this response was indeed mediated by NO. More recently, it has been shown that L-NMMA administered intracerebrally in mice inhibits the increase in cyclic GMP induced by NMDA, quisqualate, kainate, harmaline and pentylenetetrazole (Wood et al. 1990). Furthermore, the increases in cyclic GMP induced by kainate in rat cerebellar slices were also inhibited by L-NMMA (Garthwaite et al. 1989b).

Interestingly, physiological Ca^{2+} levels, which are essential for the action of NO synthase, were found to inhibit the brain soluble guanylate cyclase (Knowles et al. 1989; Olson et al. 1976). This led us to suggest that this could represent a control mechanism whereby guanylate cyclase is not activated in those cells stimulated to produce NO but only in the effector cells (Knowles et al. 1989).

The NO synthase in the brain has been further characterized and shown to be inhibited competitively by L-NMMA, L-NNA and L-NIO (Knowles et al. 1990a). Like the endothelial cell and platelet enzyme, brain NO synthase only required NADPH as a cofactor. Bovine brain cytosol has also been shown to contain NO synthase (Schmidt et al. 1989a). This enzyme has now been purified from rat cerebellum and shown to be calmodulin dependent (Bredt and Snyder 1990). The purified enzyme migrates as a single 150-kDa band on SDS/PAGE and appears to be a monomer. The biological consequences of stimulating the soluble guanylate cyclase in different parts of the brain are yet to be elucidated. NO is also released in the peripheral nervous system, but it remains to be determined whether it acts as a transmitter in these nerves or modulates the release or action of another transmitter, or both.

An L-arginine:NO synthase is present in both the cortex and the medulla of the adrenal gland (Palacios et al. 1989). This enzyme is NADPH and Ca^{2+} dependent, like that in the vascular endothelium, the platelet and the brain. The functional importance of this pathway in regulating adrenal cortex and medulla function is not clear, although cyclic GMP has been implicated in both catecholamine secretion (Derome et al. 1981; Dohi et al. 1983; O'Sullivan and Burgoyne 1990) and steroidogenesis (Perchellet et al. 1978).

Evidence for the presence of the L-arginine:NO pathway has also been found in some cell lines including mouse neuroblastoma cells (NIE-115) (Ishii et al. 1989; Gorsky et al. 1990) and porcine kidney epithelial cells (LLC-PK_1) stimulated with oxytocin (Ishii et al. 1989) or with vasopressin (Schroder and Schror 1989). Recently, rat mast cells have also been shown to produce an NO-like substance which modulates the release of histamine (Salvemini et al. 1990).

The Inducible Nitric Oxide Synthase

The Macrophage

A correlation betwen immunostimulation and elevated NO_3^- synthesis has been demonstrated in animals and man (Hegesh and Shiloah 1982; Wagner et al. 1983, 1984). Moreover, the formation of NO_2^- and NO_3^- which occurs after activation of the macrophage cell line RAW 264.7 with lipopolysaccharide (LPS) and γ-interferon (IFN-γ) was found to be dependent on the presence of L-arginine (Iyengar et al. 1987). The NO_2^- and NO_3^- formed are derived from the terminal guanidino nitrogen atom(s) of L-arginine. This reaction, which is independent of the respiratory burst, results in the formation of L-citrulline as a co-product (Iyengar et al. 1987).

An L-arginine-dependent pathway has been shown to be responsible for the cytotoxic activities of macrophages (Hibbs et al. 1987). These cytotoxic activities, which include inhibition of mitochondrial respiration, aconitase activity and DNA synthesis, are thought to be mediated by inhibition of iron-containing enzymes in target cells. These activities, as well as the generation of NO_2^- and NO_3^-, are inhibited by L-NMMA (Hibbs et al. 1987). Tumour necrosis factor increases macrophage NO_3^- production in response to IFN-γ, and this effect is also inhibited by L-NMMA (Ding et al. 1988).

Following the demonstration of the synthesis of NO from L-arginine by mammalian cells (Palmer et al. 1988a), three groups demonstrated that NO synthesized from L-arginine was the precursor of the NO_2^- and NO_3^- in the macrophage (Marletta et al. 1988; Hibbs et al. 1988; Stuehr et al. 1989). The cytotoxic properties of NO were also demonstrated, for treatment of guinea-pig hepatoma cells with NO gas under anaerobic culture conditions was shown to cause intracellular iron loss, inhibition of mitochondrial respiration, inhibition of aconitase, inhibition of DNA synthesis and cytostasis in the tumour cells (Hibbs et al. 1988; Stuehr and Nathan 1989).

The NO synthase in the macrophage differs from that in the endothelial cell, platelet and brain in that it was not detectable in either macrophage cell lines or freshly elicited macrophages that had not been activated by an agent such as LPS, alone or in combination with IFN-γ (Stuehr and Marletta 1985, 1987a, 1987b) and it required protein synthesis for its expression (Marletta et al. 1988). Studies of the NO synthase in LPS/IFN-γ-activated RAW 264.7 cells showed that the enzymatic activity (which was not present in non-activated cells) was cytosolic and was present in the $10\,000 \times g$ supernatant (Marletta et al. 1988). The enzyme required L-arginine and NADPH, but not Ca^{2+}, and its activity was enhanced by Mg^{2+} although this cation was not essential for the formation of NO. Similar results have been obtained in cytosols of the murine macrophage cell line J774 activated with LPS and IFN-γ (McCall et al. 1991). NO generation from activated macrophage cytosol has also been shown to be dependent on the presence of tetrahydrobiopterin (Tayeh and Marletta 1989; Kwon et al. 1989) and to be stimulated by flavin adenine dinucleotide and reduced glutathione (Stuehr et al. 1990).

Formation of NO_2^- and NO_3^- in the macrophage is inhibited not only by L-NMMA, but also by L-canavanine (Iyengar et al. 1987). This is a further difference between the NO synthase in the macrophage and that in the endothelial cell, platelet and brain. More recently, L-NIO has been shown to be a potent, rapid in onset, and irreversible inhibitor of NO generation in activated J774 cells (McCall et al. 1991). In contrast, L-NMMA was found to be slower in onset and L-NAME and L-NNA were considerably less potent and were fully reversible (McCall et al. 1991). These properties make L-NIO a potentially useful tool to investigate the generation of NO in phagocytic cells.

The L-arginine:NO pathway has been proposed to be a primary defence mechanism against intracellular microorganisms as well as pathogens such as fungi and helminths that are too large to be phagocytosed (Hibbs et al. 1990). In some cells NO is cytotoxic and in others cytostatic (Hibbs et al. 1990), suggesting that the sensitivity to NO varies from one cell to another. The reasons for this are not yet clear but they may be dependent on the relative importance of iron–sulfur-centered enzymes in different cells.

The Neutrophil

In 1988 glycogen-elicited rat peritoneal neutrophils were found to release a factor that relaxed vascular smooth muscle (Rimele et al. 1988) and increased cyclic GMP levels in this tissue (Lee et al. 1988). Human neutrophils were subsequently shown to inhibit platelet aggregation (Cynk et al. 1988), an effect accompanied by increased platelet cyclic GMP levels (Salvemini et al. 1989).

The release of this anti-aggregating activity from rat peritoneal and human neutrophils was inhibited by L-NMMA and by L-canavanine (McCall et al. 1989; Schmidt et al. 1989b). The identity of this EDRF-like factor released from stimulated human neutrophils was confirmed as NO using chemiluminescence (Schmidt et al. 1989b; Wright et al. 1989).

The release of NO from rat peritoneal neutrophils could also be inhibited by L-NMO (McCall et al. 1991). Unlike the effect of L-NMMA and L-canavanine, the effect of L-NIO was rapid in onset, reaching full inhibitory effect within 10 minutes, an observation similar to that in J774 cells (McCall et al. 1991). Furthermore, the maximum degree of inhibition of NO formation observed with L-NIO was significantly greater than that observed with L-NMMA. The presence of NO synthase in these cells has recently been confirmed (McCall et al. 1991). The enzyme has the same characteristics as that in the J774 cells in that it is cytosolic and Ca^{2+} independent. Interestingly, L-NAME, which is a potent inhibitor of endothelial NO synthesis (Rees et al. 1990a, 1990b), did not affect the generation of NO by the neutrophil (McCall et al. 1990, 1991). However, both L-NNA and L-NAME were weak inhibitors of NO synthase isolated from the neutrophil (McCall et al. 1991).

The biological significance of NO production by neturophils and whether the NO synthase is induced in these cells during differentiation remains to be elucidated.

Other Cells and Tissues

Evidence for an inducible NO synthase, similar to that in the macrophage, has been found in rat Kupffer cells which, when co-cultured with hepatocytes and stimulated with LPS, induced a significant suppression of hepatocyte total protein synthesis, but only when L-arginine was present in the medium (Billiar et al. 1989). This effect, which required an induction period of several hours, was associated with formation of NO_2^-, NO_3^- and citrulline both in the Kupffer cells and the hepatocytes. It was subsequently shown that the supernatant from activated Kupffer cells induced the formation of NO in the hepatocytes, an effect which was blocked by L-NMMA (Curran et al. 1989). These data have led to the suggestion that Kupffer cells activated by septic stimuli or other inflammatory states respond by forming NO themselves and by inducing NO formation in neighboring hepatocytes. A major effect of this NO is cytotoxicity since it suppresses hepatocyte protein synthesis (Curran et al. 1989). Induction by endotoxin of NO synthase in hepatocytes and in the lung has also been demonstrated in vivo (Knowles et al. 1990b).

Until recently, it was considered that the only NO synthase in the vessel wall was the constitutive, Ca^{2+}-dependent enzyme in the endothelium. However, porcine endothelial cells in culture have now been shown to express a Ca^{2+}-independent NO synthase following activation in vitro with LPS and IFN-γ (Radomski et al. 1990b). Furthermore, rings of rat aorta, with or without endothelium, incubated with LPS in vitro (Rees et al. 1990c) or obtained from rats treated with LPS (Knowles et al. 1990c) also exhibited Ca^{2+}-independent synthesis of NO. The induction of NO synthase in both the vascular endothelium and the vascular smooth muscle layer was time dependent and inhibited by cycloheximide, indicating that the enzymes were synthesized de novo and are similar to the NO synthase induced in other cells (Radomski et al. 1990b; Rees et al. 1990c). This induction of NO synthase was accompanied by loss in tone, increased levels of cyclic GMP, a decreased response to the constrictor effect of phenylephrine, and the ability of the tissues to relax to L-arginine, all of which could be prevented or reversed by inhibitors of NO synthase. The induction of the enzyme and the accompanying functional and biochemical changes were prevented by cycloheximide and by polymyxin B, an antagonist of LPS, indicating that the LPS shown to be present in the experimental system was responsible for the induction of this enzyme (Rees et al. 1990c).

The induction of an NO synthase in the vessel wall explains some observations made over the years in vascular preparations, including the hyporeactivity to vasoconstrictors (Parratt 1973; Wakabayashi et al. 1987; McKenna 1990) and the elevations in cyclic GMP in smooth muscle after incubation with LPS (Fleming et al. 1990).

An inducible NO synthase has also been identified in some non-phagocytic cell lines. EMT-6 cells, a spontaneous murine mammary adenocarcinoma cell line, when exposed to medium conditioned by activated macrophages synthesize NO_2^-, NO_3^- and L-citrulline (Amber et al. 1988). This effect is dependent on the presence of L-arginine and is accompanied by inhibition of aconitase and of DNA replication, together with the release of iron from the cells. Cycloheximide prevents all these responses, showing that protein synthesis is required for the NO synthase to be induced (Amber et al. 1988). The same pattern of response can be induced in EMT-6 cells by IFN-γ in combination with LPS, tumour necrosis factor (TNF) or interleukin-1 (IL-1), where it is associated with prevention of cell growth (Lepoivre et al. 1989). Treatment of EMT-6 cells with IFN-γ and LPS stopped not only their own proliferation but also inhibited DNA synthesis in other mouse, rat and human tumor cell lines (Lepoivre et al. 1990). Murine fibroblasts stimulated with IFN-γ + IL-1, or IFN-γ + TNFα or IFN-γ + IL-1 + TNFα have also been shown to generate NO (Hibbs et al., personal communication). These observations provide further support for the role of NO as mediator of the cytostatic actions of lymphokines.

Immunologically Induced Formation of Nitric Oxide In Vivo

Treatment of rats for seven days with *Corynbacterium parvum* induced NO synthase in hepatocytes; this correlated with an increase in blood levels of NO_2^- and NO_3^- which could be inhibited with L-NMMA (Billiar et al. 1990a). Furthermore, treatment of rats with LPS induced after 6 hours an NO synthase in the lung and liver (Knowles et al. 1990b). This NO synthase, which was not present in the tissues from untreated rats, was Ca^{2+}-independent. Moreover, in the liver the highest activity of this enzyme was localized in the hepatocytes (Knowles et al. 1990b). The induction of NO synthase in the liver was maximal at 6 hours and declined towards control levels in the next 18 hours.

The induction of a Ca^{2+}-independent NO synthase in the endothelial and smooth muscle layer of aortae from rats treated with LPS has also been demonstrated (Knowles et al. 1990c). Furthermore, the highest activity of the enzyme found was localized in the vascular smooth muscle layer.

In a model of sepsis-related hepatic injury L-NMMA was found to elevate serum levels of aspartate aminotransferase and lactate dehydrogenase, indicating increased liver damage (Billiar et al. 1990b). Further research will have to reconcile these observations with in vitro data suggesting that NO has a cytotoxic effect in hepatocytes (Curran et al. 1989). L-NMMA has been shown to potentiate LPS-induced gastrointestinal damage (Hutcheson et al. 1990) and the lethal effect of LPS in anesthetized rabbits (Moncada, Palmer, Rees and Wright, unpublished observations). In contrast, L-NMMA reverses TNF-induced hypotension in dogs (Kilbourn et al. 1990) and inhibits LPS-induced hypotension in rats (Thiemermann and Vane 1990).

It is likely, therefore, that the hypotension of endotoxemia or that induced by cytokines is mediated, at least in part, by the induction of NO synthase in the vasculature. Furthermore, a low level of induction of NO synthase may lead to vasodilatation or hypotension, with or without tissue damage, dependent on the rate of NO release. At this stage the consequences of induction of NO synthase in a variety of tissues are not yet fully understood. It is likely that in vivo, besides being cytotoxic, NO acts as a protective mechanism because of its powerful vasodilator action, an effect which cannot be observed in isolated cells or tissues. However, the interplay between these two properties of NO is not yet clearly understood and the net outcome of inhibiting NO generation is still uncertain. In view of these possibilities and apparent contradictory results the use of L-NMMA in humans for the treatment of septic shock should await the results of further research.

Inhibition by Glucocorticoids of Immunologically Induced Formation of Nitric Oxide

Glucocorticoids, which are beneficial in the prevention or treatment of endotoxin shock in animals and man (Hinshaw et al. 1980; Greisman 1982; Nicholson 1982; Hoffman et al. 1984), inhibit the induction of several proteins (for review see Haynes and Murad 1985). This led to the study of these compounds as inhibitors of the induction of NO synthase.

The glucocorticoids dexamethasone, hydrocortisone and cortisol inhibited the induction but not the activity of Ca^{2+}-independent NO synthase in vitro in vascular endothelial cells (Radomski et al. 1990b), fresh vascular tissue (Rees et al. 1990c), the macrophage cell line J774 (Di Rosa et al. 1990) and EMT-6 adenocarcinoma cells (O'Connor and Moncada, unpublished observations) after stimulation with LPS, either alone or in combination with IFN-γ. In addition, the in vitro inhibition of NO synthase by glucocorticoids was prevented by cortexolone (Radomski et al. 1990b; Di Rosa et al. 1990), a partial agonist on glucocorticoid receptors, indicating that this action is specific and therefore related to the pharmacological and probably some of the physiological effects of glucocorticoids. Furthermore, in vivo induction of NO synthase in liver, lung and vascular tissue of rats after treatment with LPS was prevented by dexamethasone (Knowles et al. 1990c) and cortisol. This action of glucocorticoids, which occurs at the low concentrations achieved in plasma during the therapeutic use of these compounds, correlates with their anti-inflammatory potency and is not shared by progesterone.

The induction by LPS of NO synthase in vascular tissue in vitro and the accompanying vascular relaxation, the hyporesponsiveness to vasoconstrictors and the increase in cyclic GMP were inhibited by incubation with dexamethasone (Rees et al. 1990c). This suggests that immunologically induced release of NO may explain at least some of the pathophysiological features of endotoxin shock.

Conclusions

The discovery of the generation of NO by mammalian tissues and the elucidation of some of its biological roles has, in the last few years, thrown new light onto many areas of research. These range from the mediation of endothelium-dependent relaxation as a general adaptive mechanism of the cardiovascular system, to the role as the transduction mechanism for the stimulation of soluble guanylate cyclase in the central nervous system and many other tissues (Moncada et al. 1989).

Immunologically released NO, in addition to being cytostatic or cytotoxic for invading microorganisms or tumor cells, may also have other functions including the mediation of pathological increases in blood flow, modulation of platelet and white cell behavior and other aspects of the immunological response. It is likely that the inducible enzyme, which was originally thought to be present only in phagocytic cells, is more widely distributed and plays a role as a general defence mechanism. The finding that glucocorticoids inhibit the induction of this NO synthase is of special significance for it may explain, at least in part, the therapeutic and toxic actions of these compounds and reveal the importance of NO in a variety of conditions.

The implications of the synthesis of NO in terms of potential novel treatments for different diseases needs to be analyzed and developed. It can, however, be predicted that, as with other fundamental biological discoveries, this will find its application in novel therapies.

References

Aisaka K, Gross SS, Griffith OW, Levi R (1989) N^G-Methylarginine, an inhibitor of endothelium-derived nitric oxide synthesis, is a potent pressor agent in the guinea pig: does nitric oxide regulate blood pressure in vivo? Biochem Biophys Res Commun 160:881–886

Amber IJ, Hibbs JB Jr, Taintor RR, Vavrin Z (1988) The L-arginine dependent effector mechanism is induced in murine adenocarcinoma cells by culture supernatant from cytotoxic activated macrophages. J Leukocyte Biol 43:187–192

Amezcua JL, Dusting GJ, Palmer RMJ, Moncada S (1988) Acetylcholine induces vasodilatation in the rabbit isolated heart through the release of nitric oxide, the endogenous nitrovasodilator. Br J Pharmacol 95:830–834

Amezcua JL, Palmer RMJ, de Souza BM, Moncada S (1989) Nitric oxide synthesized from L-arginine regulates vascular tone in the coronary circulation of the rabbit. Br J Pharmacol 97:1119–1124

Ashley RH, Brammer MJ, Marchbanks R (1984) Measurement of intrasynaptosomal free calcium by using the fluorescent indicator quin-2. Biochem J 219:149–158

Azuma H, Ishikawa M, Sekizaki S (1986) Endothelium-dependent inhibition of platelet aggregation. Br J Pharmacol 88:411–415

Billiar TR, Curran RD, Stuehr DJ, West MA, Bentz BG, Simmons RL (1989) An L-arginine dependent mechanism mediates Kupffer cell inhibition of hepatocyte protein synthesis in vitro. J Exp Med 169:1467–1472

Billiar TR, Curran RD, Stuehr DJ, Stadler J, Simmons RL, Murray SA (1990a) Inducible cytosolic enzyme activity for the production of nitrogen oxides from L-arginine in hepatocytes. Biochem Biophys Res Commun 168:1034–1040

Billiar TR, Curran RD, Stuehr DJ, Hofmann K, Simmons RL (1990b) Inhibition of L-arginine metabolism by N^G-monomethyl-L-arginine in vivo promotes hepatic damage in response to lipopolysaccharide. In: Moncada S, Higgs EA (eds) Nitric oxide from L-arginine: a bioregulatory system. Elsevier, Amsterdam, pp 275–280

Bredt DS, Snyder SH (1989) Nitric oxide mediates glutamate-linked enhancement of cGMP levels in the cerebellum. Proc Natl Acad Sci USA 86:9030–9033

Bredt DS, Snyder SH (1990) Isolation of nitric oxide synthetase, a calmodulin-requiring enzyme. Proc Natl Acad Sci USA 87:682–685

Broten TP, Miyashiro JK, Feigl EO, Moncada S (1991) The role of EDRF in parasympathetic cholinergic coronary vasodilatation. Circulation (submitted)

Chen WZ, Palmer RMJ, Moncada S (1989) The release of nitric oxide from the rabbit aorta. J Vasc Med Biol 1:2–6

Chu A, Lin C-C, Chambers DE, Kuehl WD, Palmer RMJ, Moncada S, Cobb FR (1991) Effects of inhibition of nitric oxide formation on basal tone and endothelium-dependent responses of the coronary arteries in awake dogs. J Clin Invest 87:1964–1968

Curran RD, Billar TR, Stuehr DJ, Hofmann K, Simmons RL (1989) Hepatocytes produce nitrogen oxides from L-arginine in response to inflammatory products from Kupffer cells. J Exp Med 170:1769–1774

Cynk E, Kondo K, Salvemini D, Sneddon JM, Vane JR (1988) Human nomocytes and neutrophils inhibit platelet aggregation by releasing an EDRF-like factor. J Physiol 407:23P

Deguchi T (1977) Endogenous activating factor for guanylate cyclase in synaptosomal-soluble fraction of rat brain. J Biol Chem 252:7617–7619

Deguchi T, Yoshioka M (1982) L-Arginine identifed as an endogenous activator for soluble guanylate cyclase from neuroblastoma cells. J Biol Chem 257:10147–10152

Derome G, Tseng R, Mercier P, Lemaire I, Lemaire S (1981) Possible muscarinic regulation of catecholamine secretion mediated by cyclic GMP in isolated bovine adrenal chromaffin cells. Biochem Pharmacol 30:855–860

Ding AH, Nathan CF, Stuehr DJ (1988) Release of reactive nitrogen intermediates and reactive oxygen intermediates from mouse peritoneal macrophages. J Immunol 141:2407–2412

Di Rosa M, Radomski M, Carnuccio R, Moncada S (1990) Glucocorticoids inhibit the induction of nitric oxide synthase in macrophages. Biochem Biophys Res Commun 172:1246–1252

Dohi T, Morita K, Tsujimoto A (1983) Effect of sodium azide on catecholamine release from isolated adrenal gland and on guanylate cyclase. Eur J Pharmacol 94:331–335

Fleming I, Gray GA, Julou-Schaeffer G, Parratt JR, Stoclet J-C (1990) Incubation with endotoxin activates the L-arginine pathway in vascular tissue. Biochem Biophys Res Commun 171:562–568

Fukuto JM, Wood KS, Byrns RE, Ignarro LJ (1990) N^G-Amino-L-arginine: a new potent antagonist of L-arginine-mediated endothelium-dependent relaxation. Biochem Biophys Res Commun 168:458–465

Furchgott RF (1984) The role of endothelium in the responses of vascular smooth muscle to drugs. Ann Rev Pharmacol Toxicol 24:175–197

Gardiner SM, Compton AM, Bennett T, Palmer RMJ, Moncada S (1990a) Regional haemodynamic effects of inhibiting endothelial cell production of nitric oxide with N^G-monomethyl-L-arginine in conscious rats. In: Moncada S, Higgs EA (eds) Nitric oxide from L-arginine: a bioregulatory system. Elsevier, Amsterdam, pp 81–88

Gardiner SM, Compton AM, Bennett T, Palmer RMJ, Moncada S (1990b) Control of regional blood flow by endothelium-derived nitric oxide. Hypertension 15:486–492

Gardiner SM, Compton AM, Bennett T, Palmer RMJ, Moncada S (1990c) Persistent haemodynamic changes following prolonged infusions of N^G-monomethyl-L-arginine (L-NMMA) in conscious rats. In: Moncada S, Higgs EA (eds) Nitric oxide from L-arginine: a bioregulatory system. Elsevier, Amsterdam, pp 489–491

Gardiner SM, Compton AM, Bennett T, Palmer RMJ, Moncada S (1990d) Regional haemodynamic changes during oral ingestion of N^G-monomethyl-L-arginine or N^G-nitro-L-arginine methyl ester in conscious Brattleboro rats. Br J Pharmacol 101:10–12

Garthwaite J, Charles SL, Chess-Williams R (1988) Endothelium-derived relaxing factor release on activation of NMDA receptors suggests role as intercellular messenger in the brain. Nature 336:385–388

Garthwaite J, Garthwaite G, Palmer RMJ, Moncada S (1989a) NMDA receptor activation induces nitric oxide synthesis from arginine in rat brain slices. Eur J Pharmacol 172:413–416

Garthwaite J, Southam E, Anderton M (1989b) A kainate receptor linked to nitric oxide synthesis
    from arginine. J Neurochem 53:1952–1954
Gorsky LD, Forstermann Y, Ishii K, Murad F (1990) Production of an EDRF-like activity in the
    cytosol of N1E-115 neuroblastoma cells. FASEB J 4:1494–1500
Greisman SE (1982) Experimental Gram-negative bacterial sepsis: optimal methyl-prednisolone
    requirements for prevention of mortality not preventable by antibiotics alone. Proc Soc Exp Biol
    Med 170:436–442
Haynes RC Jr, Murad F (1985) Adrenocorticotropic hormone; adrenocortical steroids and their
    synthetic analogues; inhibitors of adrenocortical steroid biosynthesis. In: Gilman A, Goodman
    LS, Rall TW, Murad F (eds) The pharmacological basis of therapeutics. Macmillan, New York,
    7th edn, pp 1459–1489
Hegesh E, Shiloah J (1982) Blood nitrates and infantile methemoglobinemia. Clin Chim Acta
    125:107–115
Hibbs JB Jr, Vavrin Z, Taintor RR (1987) L-Arginine is required for expression of the activated
    macrophage effector mechanism causing selective metabolic inhibition in target cells. J Immunol
    138:550–565
Hibbs JB Jr, Taintor RR, Vavrin Z, Rachlin EM (1988) Nitric oxide: a cytotoxic activated
    macrophage effector molecule. Biochem Biophys Res Commun 157:87–94
Hibbs JB Jr, Taintor RR, Vavrin Z, Granger DL, Drapier J-C, Amber IJ, Lancaster JR Jr (1990)
    Synthesis of nitric oxide from a terminal guanidino nitrogen atom of L-arginine: a molecular
    mechanism regulating cellular proliferation that targets intracellular iron. In: Moncada S, Higgs
    EA (eds) Nitric oxide from L-arginine: a bioregulatory system. Elsevier, Amsterdam, pp 189–223
Hinshaw LB, Archer LT, Beller-Todd BK, Coalson JJ, Flournoy DJ, Passey R, Benjamin B, White
    GL (1980) Survival of primates in LD_{100} septic shock following steroid/antibiotic therapy. J Surg
    Res 28:151–170
Hoffman SL, Punjabi NH, Kumala S, Moechtar MA, Pulungsih SP, Rivai AR, Rockhill RC,
    Woodward TE, Loedin AA (1984) Reduction of mortality in chloramphenicol-treated severe
    typhoid fever by high-dose dexamethasone. N Engl J Med 310:82–88
Hutcheson IR, Whittle BJR, Boughton-Smith NK (1990) Role of nitric oxide in maintaining vascular
    integrity in endotoxin-induced acute intestinal damage in the rat. Br J Pharmacol 101:815–820
Ignarro LJ, Buga GM, Wood KS, Byrns RE, Chaudhuri G (1987) Endothelium-derived relaxing
    factor produced and released from artery and vein is nitric oxide. Proc Natl Acad Sci USA
    84:9265–9269
Ishii K, Gorsky LD, Forstermann U, Murad F (1989) Endothelium-derived relaxing factor (EDRF):
    the endogenous activator of soluble guanylate cyclase in various types of cells. J Appl Cardiol
    4:505–512
Ishii K, Chang B, Kerwin JF, Huang ZJ, Murad F (1990) N^G-Nitro-L-arginine: a potent inhibitor of
    endothelium-derived relaxing factor formation. Eur J Pharmacol 176:219–223
Iyengar R, Stuehr DJ, Marletta MA (1987) Macrophage synthesis of nitrite, nitrate and
    N-nitrosamines: precursors and role of the respiratory burst. Proc Natl Acad Sci USA
    84:6369–6373
Kelm M, Schrader J (1988) Nitric oxide release from the isolated guinea pig heart. Eur J Pharmacol
    155:313–316
Kelm M, Feelisch M, Spahr R, Piper H-M, Noack E, Schrader J (1988) Quantitative and kinetic
    characterization of nitric oxide and EDRF released from cultured endothelial cells. Biochem
    Biophys Res Commun 154:236–244
Khan MT, Furchgott RF (1987) Additional evidence that endothelium-derived relaxing factor is
    nitric oxide. In: Rand MJ, Raper C (eds) Pharmacology. Elsevier, New York pp 341–344
Kilbourn RG, Gross SS, Jubran A, Adams J, Griffith OW, Levi R, Lodato RF (1990)
    N^G-Methyl-L-arginine inhibits tumor necrosis factor-induced hypotension: implications for the
    involvement of nitric oxide. Proc Natl Acad Sci USA 87:3629–3632
Knowles RG, Palacios M, Palmer RMJ, Moncada S (1989) Formation of nitric oxide from L-arginine
    in the central nervous system: a transduction mechanism for stimulation of the soluble guanylate
    cyclase. Proc Natl Acad Sci USA 86:5159–5162
Knowles RG, Palacios M, Palmer RMJ, Moncada S (1990a) Kinetic characteristics of nitric oxide
    synthase from rat brain. Biochem J 269:207–210
Knowles RG, Merrett M, Salter M, Moncada S (1990b) Differential induction of brain, lung and
    liver nitric oxide synthase by endotoxin in the rat. Biochem J 270:833–836
Knowles RG, Salter M, Brooks SL, Moncada S (1990c) Anti-inflammatory glucocorticoids inhibit
    the induction by endotoxin of nitric oxide synthase in the lung, liver and aorta of the rat. Biochem
    Biophys Res Commun 172:1042–1048

Kwon NS, Nathan CF, Stuehr DJ (1989) Reduced biopterin as a cofactor in the generation of nitrogen oxides by murine macrophages. J Biol Chem 264:20496–20501

Lee DKH, Faunce D, Henry D, Sturm R, Rimele T (1988) Rat polymorphonuclear leukocytes (PMN) increase cGMP levels in rat aorta. FASEB J 2:A518

Lepoivre M, Boudbid H, Petit J-F (1989) Antiproliferative activity of γ-interferon combined with lipopolysaccharide on murine adenocarcinoma: dependence on an L-arginine metabolism with production of nitrite and citrulline. Cancer Res 49:1970–1976

Lepoivre M, Boudbid H, Petit J-F (1990) Antitumour cytostatic effect of EMT6 cells treated with interferon-γ and lipopolysaccharide. In: Moncada S, Higgs EA (eds) Nitric oxide from L-arginine: a bioregulatory system. Elsevier, Amsterdam, pp 415–422

Marletta MA, Yoon PS, Iyengar R, Leaf CD, Wishnok JS (1988) Macrophage oxidation of L-arginine to nitrite and nitrate: nitric oxide is an intermediate. Biochemistry 27:8706–8711

Mayer B, Schmidt K, Humbert R, Bohme E (1989) Biosynthesis of endothelium-derived relaxing factor: a cytosolic enzyme in porcine aortic endothelial cells Ca^{2+}-dependently converts L-arginine into an activator of soluble guanylyl cyclase. Biochem Biophys Res Commun 164:678–685

McCall TB, Boughton-Smith NK, Palmer RMJ, Whittle BJR, Moncada S (1989) Synthesis of nitric oxide from L-arginine by neutrophils: release and interaction with superoxide anion. Biochem J 261:293–296

McCall T, Palmer RMJ, Boughton-Smith N, Whittle BJR, Moncada S (1990) The L-arginine:nitric oxide pathway in neutrophils. In: Moncada S, Higgs EA (eds) Nitric oxide from L-arginine: a bioregulatory system. Elsevier, Amsterdam, pp 257–265

McCall TB, Feelisch M, Palmer, RMJ, Moncada S (1991) Identification of N-iminoethyl-L-ornithine as an irreversible inhibitor of nitric oxide synthase in phagocytic cells. Br J Pharmacol 102:234–238

McKenna TM (1990) Prolonged exposure of rat aorta to low levels of endotoxin in vitro results in impaired contractility: association with vascular cytokine release. J Clin Invest 86:160–168

Miki N, Kawabe Y, Kuriyama K (1977) Activation of cerebral guanylate cyclase by nitric oxide. Biochem Biophys Res Commun 75:851–856

Moncada S, Palmer RMJ (1990) The L-arginine:nitric oxide pathway in the vessel wall. In: Moncada S, Higgs EA (eds) Nitric oxide from L-arginine: a bioregulatory system. Elsevier, Amsterdam, pp 19–33

Moncada S, Radomski MW, Palmer RMJ (1988a) Endothelium-derived relaxing factor: identification as nitric oxide and role in the control of vascular tone and platelet function. Biochem Pharmacol 37:2495–2501

Moncada S, Palmer RMJ, Higgs EA (1988b) The discovery of nitric oxide as the endogenous nitrovasodilator. Hypertension 12:365–372

Moncada S, Palmer RMJ, Higgs EA (1989) Biosynthesis of nitric oxide from L-arginine: a pathway for the regulation of cell function and communication. Biochem Pharmacol 38:1709–1715

Moore PK, al Swayeh OA, Chong NWS, Evans R, Mirzazadeh S, Gibson A (1989) L-N^G-Nitroarginine (NOARG) inhibits endothelium-dependent vasodilatation in the rabbit aorta and perfused rat mesentery. Br J Pharmacol 98:905P

Mulsch A, Busse R (1990) N^G-Nitro-L-arginine (N^5-[imino(nitroamino)methyl]-L-ornithine) impairs endothelium-dependent dilations by inhibiting cytosolic nitric oxide synthesis from L-arginine. Naunyn-Schmiedebergs Arch Pharmacol 341:143–147

Mulsch A, Bassenge E, Busee R (1989) Nitric oxide synthesis in endothelial cytosol: evidence for a calcium-dependent and a calcium-independent mechanism. Naunyn-Schmiedebergs Arch Pharmacol 340:767:770

Nicholson DP (1982) Glucocorticoids in the treatment of shock and the adult respiratory distress syndrome. Clin Chest Med 3:121–132

Olson DR, Kon C, Breckenridge BM (1976) Calcium ion effects on guanylate cyclase of brain. Life Sci 18:935–940

O'Sullivan AJ, Burgoyne RD (1990) Cyclic GMP regulates nicotine-induced secretion from cultured bovine adrenal chromaffin cells: effects of 8-bromo-cyclic GMP, atrial natriuretic peptide, and nitroprusside (nitric oxide). J Neurochem 54:1805–1808

Palacios M, Knowles RG, Palmer RMJ, Moncada S (1989) Nitric oxide from L-arginine stimulates the soluble guanylate cyclase in adrenal glands. Biochem Biophys Res Commun 165:802–809

Palmer RMJ, Moncada S (1989) A novel citrulline-forming enzyme implicated in the formation of nitric oxide by vascular endothelial cells. Biochem Biophys Res Commun 158:348–352

Palmer RMJ, Ferrige AG, Moncada S (1987) Nitric oxide release accounts for the biological activity of endothelium-derived relaxing factor. Nature (Lond) 327:524–526

Palmer RMJ, Ashton DS, Moncada S (1988a) Vascular endothelial cells synthesize nitric oxide from L-arginine. Nature (Lond) 333:664–666

Palmer RMJ, Rees DD, Ashton DS, Moncada S (1988b) L-Arginine is the physiological precursor for the formation of nitric oxide in endothelium-dependent relaxation. Biochem Biophys Res Commun 153:1251–1256

Parratt JR (1973) Myocardial and circulatory effects of E. coli endotoxin; modification of responses to catecholamines. Br J Pharmacol 47: 12–25

Perchellet JP, Shanker G, Sharma RK (1978) Regulatory role of guanosine 3', 5'-monophosphate in adrenocorticotropin hormone-induced steroidogenesis. Science 199:311–312

Radomski MW, Palmer RMJ, Moncada S (1987a) Comparative pharmacology of endothelium-derived relaxing factor, nitric oxide and prostacyclin in platelets. Br J Pharmacol 92:181–187

Radomski MW, Palmer RMJ, Moncada S (1987b) The role of nitric oxide and cGMP in platelet adhesion to vascular endothelium. Biochem Biophys Res Commun 148:1482–1489

Radomski MW, Palmer RMJ, Moncada S (1987c) The anti-aggregating properties of vascular endothelium: interactions between prostacyclin and nitric oxide. Br J Pharmacol 92:639–646

Radomski MW, Palmer RMJ, Moncada S (1990a) An L-arginine:nitric oxide pathway present in human platelets regulates aggregation. Proc Natl Acad Sci USA 87:5193–5197

Radomski MW, Palmer RMJ, Moncada S (1990b) Glucocorticoids inhibit the expression of an inducible, but not the constitutive, nitric oxide synthase in vascular endothelial cells. Proc Natl Acad Sci USA 87:10043–10047

Rees DD, Palmer RMJ, Hodson HF, Moncada S (1989a) A specific inhibitor of nitric oxide formation from L-arginine attenuates endothelium-dependent relaxation. Br J Pharmacol 96:418–424

Rees DD, Palmer RMJ, Moncada S (1989b) Role of endothelium-derived nitric oxide in the regulation of blood pressure. Proc Natl Acad Sci USA 86:3375–3378

Rees DD, Palmer RMJ, Schulz R, Hodson HF, Moncada S (1990a) Characterization of three inhibitors of endothelial nitric oxide synthase in vitro and in vivo. Br J Pharmacol 101:746–752

Rees DD, Schulz R, Hodson HF, Palmer RMJ, Moncada S (1990b) Identification of some novel inhibitors of the vascular nitric oxide synthase in vivo and in vitro. In: Moncada S, Higgs EA (eds) Nitric oxide from L-arginine: a bioregulatory system. Elsevier, Amsterdam, pp 485–487

Rees DD, Cellek S, Palmer RMJ, Moncada S (1990c) Dexamethasone prevents the induction by endotoxin of a nitric oxide synthase and the associated effects on vascular tone: an insight into endotoxin shock. Biochem Biophys Res Commun 173: 541–547

Rimele TJ, Sturm RJ, Adams LM, Henry DE, Heaslip RJ, Weichman BM, Grimes D (1988) Interaction of neutrophils with vascular smooth muscle: identification of a neutrophil-derived relaxing factor. J Pharmacol Exp Ther 245:102–111

Sakuma I, Stuehr D, Gross SS, Nathan C, Levi R (1988) Identification of arginine as a precursor of endothelium-derived relaxing factor (EDRF). Proc Natl Acad Sci USA 85:8664–8667

Salvemini D, De Nucci G, Gryglewski RJ, Vane JR (1989) Human neutrophils and mononuclear cells inhibit platelet aggregation by releasing a nitric oxide-like factor. Proc Natl Acad Sci USA 86:6328–6332

Salvemini D, Masini E, Anggard E, Mannaioni PF, Vane JR (1990) Synthesis of a nitric oxide-like factor from L-arginine by rat serosal mast cells: stimulation of guanylate cyclase and inhibition of platelet aggregation. Biochem Biophys Res Commun 169:596–601

Schmidt HHHW, Nau H, Wittfoht W, Gerlach J, Prescher K-E, Klein MM, Niroomand F, Bohme E (1988) Arginine is a physiological precursor of endothelium-derived nitric oxide. Eur J Pharmacol 154:213–216

Schmidt HHHW, Wilke P, Evers B, Bohme E (1989a) Enzymatic formation of nitrogen oxides from L-arginine in bovine brain cytosol. Biochem Biophys Res Commun 165:284–291

Schmidt HHHW, Seifert R, Bohme E (1989b) Formation and release of nitric oxide from human neutrophils and HL-60 cells induced by chemotactic peptide, platelet activating factor and leukotriene B_2. FEBS Lett 244:357–360

Schroder H, Schror K (1989) Cyclic GMP stimulation by vasopressin in LLC-PK$_1$ kidney epithelial cells is L-arginine-dependent. Naunyn Schmiedebergs Arch Pharmacol 340:475–477

Stuehr DJ, Marletta MA (1985) Mammalian nitrate biosynthesis: mouse macrophages produce nitrite and nitrate in response to Escherichia coli lipopolysaccharide. Proc Natl Acad Sci USA 82:7738–7742

Stuehr DJ, Marletta MA (1987a) Induction of nitrite/nitrate synthesis in murine macrophages by BCG infection, lymphokines or interferon-γ. J Immunol 139:518–525

Stuehr DJ, Marletta MA (1987b) Synthesis of nitrite and nitrate in murine macrophage cell lines. Cancer Res 47:5590–5594

Stuehr DJ, Nathan CF (1989) Nitric oxide: a macrophage product responsible for cytostasis and respiratory inhibition in tumor target cells. J Exp Med 169:1543–1555

Stuehr D, Gross S, Sakuma I, Levi R, Nathan C (1989) Activated murine macrophages secrete a metabolite of arginine with the bioactivity of endothelium-derived relaxing factor and the chemical reactivity of nitric oxide. J Exp Med 169:1011–1020

Stuehr DJ, Kwon NS, Nathan CF (1990) FAD and GSH participate in macrophage synthesis of nitric oxide. Biochem Biophys Res Commun 168:558–565

Tayeh MA, Marletta MA (1989) Macrophage oxidation of L-arginine to nitric oxide, nitrite and nitrate. Tetrahydrobiopterin is required as a cofactor. J Biol Chem 264:19654–19658

Thiemermann C, Vane J (1990) Inhibition of nitric oxide synthesis reduces the hypotension induced by bacterial lipopolysaccharides in the rat in vivo. Eur J Pharmacol 182:591–595

Tolins J, Palmer RMJ, Moncada S, Raij L (1990) Role of endothelium-derived relaxing factor in regulation of renal hemodynamic responses. Am J Physiol 258:H655–H662

Vallance P, Collier J, Moncada S (1989a) Effects of endothelium-derived nitric oxide on peripheral arteriolar tone in man. Lancet II:997–1000

Vallance P, Collier J, Moncada S (1989b) Nitric oxide synthesised from L-arginine mediates endothelium-dependent dilatation in human veins. Cardiovasc Res 23:1053–1057

Wagner DA, Young VR, Tannenbaum SR (1983) Mammalian nitrate biosynthesis: incorporation of $^{15}NH_3$ into nitrate is enhanced by endotoxin treatment. Proc Natl Acad Sci USA 80:4518–4521

Wagner DA, Young VR, Tannenbaum SR, Schultz DS, Deen WM (1984) Mammalian nitrate biochemistry: metabolism and endogenous synthesis. In: O'Neill IK, von Borstel RC, Long JE, Miller CT, Martsch H (eds) N-Nitroso compounds: occurrence, biological effects and relevance to human cancer. IARC Scientific Publication no. 57:247–253

Wakabayashi I, Hatake K, Kakashita E, Nagai K (1987) Diminution of contractile response of the aorta from endotoxin-injected rats. Eur J Pharmacol 141:117–122

Whittle BJR, Lopez-Belmonte J, Rees DD (1989) Modulation of the vasodepressor actions of acetylcholine, bradykinin, substance P and endothelin in the rat by a specific inhibitor of nitric oxide formation. Br J Pharmacol 98:646–652

Wood PL, Emmett MR, Rao TS, Cler J, Mick S, Iyengar S (1990) Inhibition of nitric oxide synthase blocks N-methyl-D-aspartate-, quisqualate-, kainate-, harmaline- and pentylenetetrazole-dependent increases in cerebellar cyclic GMP in vivo. J Neurochem 55:346–348

Wright DC, Mulsch A, Busse R, Osswald H (1989) Generation of nitric oxide by human neutrophils. Biochem Biophys Res Commun 160:813–819

Yoshikawa K, Kuriyama K (1980) Characterization of cerebellar guanylate cyclase using N-methyl-N-nitro-N-nitrosoguanidine: presence of two different types of guanylate cyclase in soluble and particulate fraction. Biochim Biophys Acta 628:377–387

Chapter 3

Endothelial Activation: Its Role in Inflammation, Vascular Injury and Atherogenesis

R. S. Cotran

Introduction

This chapter will briefly review evidence that endothelial cells respond to cytokines by undergoing a variety of functional and structural changes – collectively called endothelial activation – which may play a role in the pathogenesis of vascular injury, including atherogenesis, the major topic of this book. Endothelial injury is a component of most theories of atherogenesis, and is central to the "response to injury" hypothesis of atherosclerosis (Munro and Cotran 1988; Ross 1986). While the initial formulation of this theory presumed that denuding endothelial injury was an early and important event (Ross and Glomset 1976), it has since become clear that endothelial injury can be non-denuding, leading to so-called endothelial dysfunction (Gimbrone 1980). It has also become apparent that one of the earliest events in atherogenesis is the adhesion of leukocytes – principally monocytes – to the endothelium, followed by their emigration across the arterial wall, and transformation to foamy macrophages (reviewed by Gerrity in this volume). Because increased leukocyte adhesivity, as will presently be described, is one of the major manifestations of activated endothelium, its possible link to atherogenesis is obvious.

Studies of endothelial cells in culture, coupled with in vivo work on the microcirculation, allow us to differentiate operationally between two mechanisms by which endothelial function may be modulated (Pober and Cotran 1990a). The first, which we shall call endothelial stimulation, does not require RNA or protein synthesis, begins rapidly, usually within seconds or minutes, and is also rapidly reversible. Examples of endothelial stimulation are the acute effects of histamine in eliciting endothelial cell contraction and increased vascular permeability, and the rapid action of thrombin in causing redistribution to the endothelial surface of the adhesive molecule GMP-140 (see below).

Endothelial activation, in contrast, requires RNA and protein synthesis, is usually delayed by hours, and sometimes days, lasts for hours or days, and is induced by endotoxin, cytokines, and certain growth factors (Pober and Cotran 1990a; Cotran and Pober 1990).

Although the term "endothelial activation" was initially used in the 1960s to describe morphological changes of hypertrophy in endothelial cells in certain immunological reactions, such as the delayed hypersensitivity reaction (Willms-Kretschmer et al. 1967), the two sets of studies which defined it in modern cell biological terms were done on endothelial cultures in the mid-1980s. In the first it was shown that interleukin-1 (IL-1), tumour necrosis factor (TNF), and endotixin induced the synthesis and surface expression of procoagulant tissue factor-like activity in cultured human umbilical vein endothelial cells (Bevilacqua et al. 1984, 1986). A large number of subsequent studies (reviewed in Pober and Cotran 1990b) then showed that these cytokines induce a variety of other alterations in endothelial-derived procoagulant and fibrinolytic properties, rendering the endothelial surface thrombogenic. The second set of studies showed that IL-1, TNF and endotoxin rendered cultured endothelial cells more adhesive to leukocytes, neutrophils, monocytes, lymphocytes, basophils and eosinophils (Bevilacqua et al. 1985). This increased adhesivity was dependent upon RNA and protein synthesis, and studies by several groups have subsequently shown that increased adhesion was dependent upon the induction and surface expression of endothelial adhesion molecules which interacted with complementary receptors on leukocytes to enhance adhesion. There are currently three well-characterized endothelial adhesion molecules.

1. *Endothelial–leukocyte adhesion molecule-1 (ELAM-1)*. This molecule is a 115-kDa single-chain glycoprotein, not normally present on endothelium, but which can be induced by IL-1, TNF and endotoxin. Its surface expression peaks 4–6 hours after stimulation, at the height of increased adhesion of neutrophils and certain leukocyte cell lines, such as HL-60. Structurally, ELAM-1 is characterized by a lectin-like domain at the amino acid terminus, a domain bearing homology to epidermal growth factor, and a tandem array of domains with consensus repeats similar to those of complement regulatory proteins (Bevilacqua et al. 1989). This structure is shared by two other adhesion proteins, GMP-140 (granule membrane protein 140) and MEL-14/Leu-8, and the family of these adhesion proteins is currently called selectin or LEC-CAM) (Osborn 1990; Brandley et al. 1990). GMP-140 is normally present in the alpha granules of platelets and in the Weibel–Palade granules of endothelial cells. Upon stimulation with histamine or thrombin, the molecule is rapidly redistributed and appears on the surface of these cells, mediating increased adhesion of neutrophils to platelets and endothelium (Johnston et al. 1989; McEver et al. 1989). Mel-14/Leu-8 is present on lymphocytes and serves as a homing receptor to the high-endothelial venules of lymphoid organs (Camerini et al. 1989). The neutrophil receptor which binds to ELAM-1 has recently been shown to be a sialyated form of the Lewix molecule (S. Le^x), or sialyated CD-15 (Walz et al. 1990).

2. *Intracellular adhesion molecule-1 (ICAM-1)*. This is a 80–90-kDa single-chain glycoprotein, normally present on endothelial cells, lymphocytes and fibroblasts, but whose basal expression on endothelial cells is increased with the cytokines IL-1 and TNF (Dustin and Springer 1988; Pober et al. 1986). The

time sequence of stimulation differs from that of ELAM-1 in that it is more delayed, plateauing at 24 hours. The receptor for ICAM-1 on leukocytes is the integrin CD-11a/CD-18 also known as lymphocyte function-associated antigen (LFA-1) (Dustin and Springer 1988). ICAM-1 has been implicated in cytokine-stimulated adhesion of neutrophils and lymphocytes. Structurally, ICAM-1 is a member of the immunoglobulin gene superfamily (Pober et al. 1986).

3. *INCAM-110/VCAM-1*. This is a 100-kDa protein, inducible by IL-1, TNF and IL-4, which mediates the binding of monocytes, lymphocytes and certain melanoma cell lines (Rice and Bevliacqua 1989; Rice et al. 1990; Osborn et al. 1989). Structurally, VCAM-1 is a member of the immunoglobulin gene superfamily and its receptor on leukocytes is the leukocyte integrin VLA (Elices et al. 1990). VCAM-1 is also present on other cell types, including follicular dendritic cells of lymph nodes, and certain epithelia (Rice and Bevilacqua 1989).

In addition to their effects on coagulation and leukocyte adhesion, cytokines induce a number of other actions in cultured endothelial cells (Cotran and Pober 1990; Pober and Cotran 1990b). They increase prostacyclin synthesis, stimulate PDGF secretion, and cause a reorganization of endothelial cell morphology from the strictly contact inhibited and normally polygonal endothelial cell, to a more fibroblastoid-cell shape. Of interest is that IL-1 and TNF both cause stimulation of cytokine secretion by the endothelial cells themselves. The array of cytokines that can thus be produced by endothelium include IL-1, IL-6, GM-CSF, G-CSF, IL-8 and MCP-1 (Cotran and Pober 1990; Pober and Cotran 1990b). This endogenous production of endothelial cytokines is important to remember in the context of more chronic lesions, such as those of atherosclerosis, as it may account for secondary cytokine effects far removed from the primary initiating stimulus or event.

Another cytokine that can prominently activate endothelial cells is IFN-γ. As detailed elsewhere (Pober and Cotran 1990b; Pober et al. 1990), IFN-γ causes induction of class II MHC molecules by endothelial cells, increased expression of class I MHC molecules, increased expression of ICAM-1, increased lymphocyte adhesion, a number of morphological changes, inhibition of cell growth and enhancement of TNF-induced activation.

Endothelial Activation In Vivo

There is ample evidence that the process of endothelial activation, discovered from in vitro work, occurs and can be mimicked in vivo. In the first place, activation antigens can be induced in experimental animals by injections of cytokines or in conditions in which cytokines are known to be produced. In a series of studies in the skin of baboons, we have shown that IL-1, TNF and endotoxin cause early (by 2 hours) adhesion and infiltration of neutrophils associated with the appearance of ELAM-1 expression on the endothelial surface of venules (Munro et al. 1989, 1991). In the case of IL-1 and TNF in subcutaneous injections, this is followed, beginning 6–9 hours after injections with adhesion and influx of mononuclear cells, associated temporally with increased expression of ICAM-1 in venular endothelium (Munro et al. 1989). In baboons subjected to septic shock, there is extensive endothelial-specific expression of ELAM-1 in the endothelium of the lung, liver, skin, adrenal gland, connective tissue, glomerular and peritubular capillaries, and also arteries, arterioles and veins (Cotran and Pober 1990; Redl et al. 1991). In certain but not all organs, ELAM-1 expression is associated with adherent neutrophils. Such ELAM-1 expression is consistent with direct effects of endotoxin, and also with the very high levels of circulating TNF in such animals. In contrast, ELAM-1 staining is much less pronounced in baboons with traumatic stroke – hypovolemic shock in which the circulating levels of cytokines are low (Cotran et al. 1986).

Secondly, endothelial activation antigens can also be induced in humans, either by the injection of recombinant cytokines, or by inducing lesions in which local cytokines are known to be produced. Induction of delayed hypersensitivity reaction in the skin of human volunteers elicits ELAM-1 (Cotran et al. 1986) and VCAM-1 (Rice et al. 1991) expression at the height of the reaction when activated lymphocytes and macrophages, known to secrete the cytokines, are abundant (Rice et al. 1991). In the late-phase reaction of immediate IgE-mediated hypersensitivity, intracutaneous injection of allergen (ragweed) in atopic subjects induces ELAM-1 expression in the endothelium very early (about 2 hours in the reaction) (Leung et al. 1991). Such ELAM-1 expression can also be induced upon in vitro incubation of atopic skin with allergen, is independent of circulating cells, and is dependent upon the local secretion of IL-1 and TNF I-cells endogenous to the skin (?mast cells) (Leung et al. 1991; Klein et al. 1989). In another study (Cotran et al. 1988), we have demonstrated that patients treated with IL-2 for advanced cancer also develop widespread endothelial activation, as reflected by ELAM-1 endothelial straining, and increased ICAM-1 and HLA-DR, possibly related to the endogenous production of cytokines (IL-1, TNF, IFN-γ) by the IL-2.

The third evidence for the in vivo relevance of endothelial activation is the frequent presence, as judged by immunohistochemical studies, of endothelial activation antigens in inflammatory, immune and neoplastic conditions associated with local production of cytokines. These include acute appendicitis, delayed hypersensitivity reaction, Hodgkin's disease, T-cell lymphoma, allograft rejection, vasculitis and a variety of cutaneous lesions (Cotran and Pober 1988). In the latter, the lesions are associated with an active inflammatory infiltrate and in most they are immunologically mediated.

Endothelial Activation and Endothelial Injury

Although endothelial activation represents a functional alteration in endo-thelium that can be differentiated from endothelial injury, there are several mechanisms by which cytokine activation of endothelium may result in increased endothelial permeability, endothelial cell lysis or endothelial cell detachment. These may be due to the direct effect of the cytokines on endothelium, or may be mediated by leukocytes (which themselves can also be activated by cytokines). There is in vivo and in vitro evidence that TNF increases endothelial permeability directly, without leukocyte participation. For example, TNF in vitro causes cytoskeletal reorganization and endothelial cell retraction, resulting in increased permeability of endothelial cell monolayers by a mechanism involving a pertussis toxin-sensitive G-protein. In vivo, severely neutropenic animals develop similar increases in lung permeability to TNF as those with normal white cell counts (Horvath et al. 1988). On the other hand, TNF also induces leukocyte aggregation and activation, and through the release of oxygen-free radicals or proteolytic enzymes may cause or enhance endothelial cell lysis or detachment (Varani et al. 1989). In addition there is evidence that cytokine-treated endothelial cells are more susceptible to lysis by neutrophils and also more susceptible to non-neutrophil-dependent endothelial injury (Ward and Varani 1990).

An altogether different mechanism of endothelial lysis related to endothelial activation is suggested by studies on the vasculitis of Kawasaki disease. The etiology of this febrile illness is unknown, but children with Kawasaki disease have high levels of circulating TNF and IFN-γ, as well as evidence of both T-cell and polyclonal B-cell activation. During the acute phase of the disease, children are shown to have circulating antibodies that are capable of lysing cultured endothelial cells in culture that have been treated either with TNF or with IFN-γ, but not untreated cultured cells (Leung et al. 1986a, 1986b). The antibodies are therefore directed against some endothelial moiety which is present in cytokine-activated endothelium. The nature of the antigen is currently unknown. Skin biopsies of patients with acute Kawasaki disease also show the presence of endothelial activation antigens (such as ELAM-1) in dermal venules. Of interest is that when such patients are treated with intravenous gamma globulin, which dramatically improves the symptoms of the disease and markedly reduces the incidence of vasculitis and coronary artery aneurysms, (Leung et al. 1989), the activation antigens disappear in the skin. This is associated with a reduction of the ability of peripheral blood mononuclear cells from these patients to elaborate IL-1. It is thus possible that new antigens induced by cytokine activation in Kawasaki disease may lead to the production of antibodies directed against these antigens. The resultant immune endothelial injury may initiate endothelial lysis as a prelude to the mural vasculitis which ensues.

Endothelial Activation and Atherogenesis

A recent study by Cybulsky and Gimbrone (1991) suggests that induction of an endothelial–leukocyte adhesion molecule may play a role in mononuclear leukocyte recruitment during atherogenesis. These authors described a monoclonal antibody to endotoxin-activated rabbit endothelial cells which significantly inhibited the adhesion of elutriated human blood monocytes and lymphocytes, as well as the leukocyte cell lines U937, HL-60 and THP-1. The molecule recognized by the antibody was eventually identified as a homologue of human INCAM-110/VCAM-1 (see above), which also mediates monocyte and lymphocyte cell adhesion. Of great interest is that whereas the molecule is absent in normal rabbit aortic endothelium, immunohistochemical studies show it to be localized to the aortic endothelium overlying early foam-cell lesions in rabbits with diet-induced hypercholesterolemia and in the Watanabe-heritable hyperlipidemic rabbit. Besides its potential pathogenetic role in inducing monocyte adhesion, this athero-ELAM may provide a molecular marker of early atherosclerosis.

Conclusions and Summary

Endothelial activation can be induced by endotoxin and cytokines, and is manifested functionally by increased adhesiveness to leukocytes, enhanced surface thrombogenicity, and by the synthesis, surface expression and/or secretion of biologically active molecules which are involved in inflammation and immunity. Evidence from animal and human studies indicates that endothelial activation occurs in vivo, after local and systemic cytokine injections and in pathological conditions associated with local or systemic cytokine production. Endothelial activation in particular appears to be involved in the pathogenesis of endotoxic shock, the vascular and cellular events of immune reactions (such as delayed hypersensitivity and allograft rejection) and certain forms of autoimmune endothelial injury. Cytokine-induced endothelial activation may also lead to endothelial injury, including increased vascular permeability, cell lysis, and detachment by direct, leucocyte-dependent and immune-mediated mechanisms. Finally, recent evidence suggests that induction of an endothelial–leukocyte adhesion molecule may be an early event in atherogenesis.

References

Bevilacqua MP, Pobert JS, Majeau GR, Cotran RS, Gimbrone MA Jr (1984) Interleukin-1 (IL-1) induces biosynthesis and cell surface expression of procoagulant activity in human vascular endothelial cells. J Exp Med 160:618–623

Bevilacqua MP, Pober JS, Wheeler ME, Cotran RS, Gimbrone MA Jr (1985) Interleukin-1 acts on cultured human vascular endothelium to increase the adhesion of polymorphonuclear leukocytes, monocytes and related leukocytic cell lines. J Clin Invest 76:2003–2011

Bevilacqua MP, Pober JS, Majeau GR, Fiers W, Cotran RS, Gimbrone MA Jr (1986) Recombinant tumor necrosis factor induces procoagulant activity in cultured human vascular endothelium: characterization and comparison with the action of interleukin-1. Proc Natl Acad Sci USA 83:4533–4537

Bevilacqua MP, Pober JS, Mendrick DL, Cotran RS, Gimbrone MA Jr (1987) Identification of an inducible endothelial–leukocyte adhesion molecule. Proc Natl Acad Sci USA 84:9238–9242

Bevilacqua MP, Stengelin S, Gimbrone MA Jr, Seed B (1989) Endothelial leukocyte adhesion molecule 1: an inducible receptor for neutrophils related to complement regulatory proteins and lectins. Science (Wash DC) 243:1160–1165

Brandley BK, Swiedler SJ, Robbins PW (1990) Carbohydrate ligands of the LEC cell adhesion molecules. Cell 63:861–863

Brett J, Gerhach H, Nawroth P, Steinberg S, Godman G, Stern D (1989) Tumor necrosis factor/cachectin increases permeability of endothelial cell monolayers by a mechanism involving regulatory G proteins. J Exp Med 169:1977–1991

Camerini D, James SP, Stamenkovic I, Seed B (1989) Leu-8/TQ1 is the human equivalent of the Mel-14 lymph node homing receptor. Nature (Lond) 342:78–82

Cotran RS, Pober JS (1988) Endothelial activation: its role in inflammatory and immune reactions. In: Simionescu N, Simionescu M (eds) Endothelial cell biology. Plenum, New York, pp 335–347

Cotran RS, Pober JS (1990) Cytokine–endothelial interactions in inflammation, immunity and vascular injury. J Am Soc Nephrol 1:225–235

Cotran RS, Gimbrone MA Jr, Bevilacqua MP, Mendrick DL, Pober JS (1986) Induction and detection of a human endothelial activation antigen in vivo. J Exp Med 164:661–666

Cotran RS, Pober JS, Gimbrone MA Jr, Springer TA, Wiebke EA, Gaspari AA, Rosenberg SA, Lotze MT (1988) Endothelial activation during interleukin 2 immunotherapy: a possible mechanism for the vascular leak syndrome. J Immunol 139:1883–1888

Cybulsky MI, Gimbrone MA Jr (1991) Localized expression of a mononuclear leukocyte adhesion molecule by vascular endothelium during atherogenesis. Science 251:788–791

Dustin ML, Springer TA (1988) Lymphocyte function-assocationed antigen-1 (LFA-1) interaction with intercellular adhesion molecule-1 (ICAM-1) is one of at least three mechanisms for lymphocyte adhesion to cultured endothelial cells. J Cell Biol 107:321–331

Elices MJ, Osborn L, Takada Y et al. (1990) VCAM-1 on activated endothelium interacts with the leukocyte integrin VLA-4 at a site distinct from the VLA-4/fibronectin binding site. Cell 60:577–584

Gimbrone MA Jr (1980) Endothelial dysfunction and the pathogenesis of atherosclerosis. In: Gotto A (ed) Atherosclerosis – V, Proceedings of the Vth International Symposium on Atherosclerosis, Springer-Verlag, New York, p 415

Horvath CJ, Ferro TJ, Jesmok G, Malik AB (1988) Recombinant tumor necrosis factor increases pulmonary vascular permeability independent of neutrophils. Proc Natl Acad Sci USA 85:9212–9223

Johnston GI, Cook RG, McEver RP (1989) Cloning of CMP-140, a granule membrane protein of platelets and endothelium: sequence similarity to proteins involved in cell adhesion and inflammation. Cell 56:1033–1044

Klein LM, Lavker RM, Matis WL, Murphy GF (1989) Degranulation of human mast cells induces an endothelial antigen central to leukocyte adhesion. Proc Natl Acad Sci USA 86:8972–8976

Leung DYM, Geha RS, Newburger JW et al. (1986a) Two monokines, interleukin-α and tumor necrosis factor, render cultured vascular endothelial cells susceptible to lysis by antibodies circulating during Kawasaki syndrome. J Exp Med 164:1958

Leung DYM, Collins T, Lapierre LA, Geha RS, Pober JS (1986b) Immunoglobulin M antibodies present in the actue phase of Kawasaki syndrome lyse cultured vascular endothelial cells stimulated by gamma interferon. J Clin Invest 77:1428

Leung DYM, Cotran RS, Kurt-Jones EA et al. (1989) Endothelial activation and high interleukin-1 secretion in the pathogenesis of acute Kawasaki disease. Lancet 2:1298

Leung DYM, Pober JS, Cotran RS (1991) Expression of endothelial leukocyte adhesion molecule-1 (ELAM-1) in elicited late-phase allergic reactions. J Clin Invest (in press)

McEver RP, Beckstead JH, Moore KL, Marshall-Carlson L, Bainton DF (1989) GMP-140, a platelet α-granule membrane protein, is also synthesized by vascular endothelial cells and is localized in Weibel–Palade bodies. J Clin Invest 84:92–99

Munro JM, Cotran RS (1988) The pathogenesis of atherosclerosis: atherogenesis and inflammation. Lab Invest 58:249–261

Munro JM, Pober JS, Cotran RS (1989) Tumor necrosis factor and interferon-γ induce distinct patterns of endothelial activation and leukocyte accumulation in skin of *Papio anubis*. Am J Pathol 133:121–133

Munro JM, Pober JS, Cotran RS (1991) Recruitment of neutrophils in the local endotoxin response: association with *de novo* endothelial expression of the adhesion molecule ELAM-1. Lab Invest 64:295–299

Osborn L (1990) Leukocyte adhesion to endothelium in inflammation. Cell 62:3–6

Osborn L, Hession C, Tizard R et al. (1989) Direct expression cloning of vascular cell adhesion molecule 1, a cytokine-induced endothelial protein that binds to lymphocytes. Cell 59:1203–1211

Pober JS, Cotran RS (1990a) The role of endothelial cells in inflammation. Transplantation 50:537–544

Pober JS, Cotran RS (1990b) Cytokines and endothelial cell biology. Phys Rev 70:427–451

Pober JS, Gimbrone MA Jr, Lapierre LA et al. (1986) Overlapping patterns of activation of human endothelial cells by interleukin-1, tumor necrosis factor and immune interferon. J Immunol 137:1893–1896

Pober JS, Doukas J, Hughes CCW, Savage COS, Munro JM, Cotran RS (1990) The potential roles of vascular endothelium in immune reactions. Hum Immunol 28:258–262

Redl H, Dinges HP, Buurman WA, van der Linden CJ, Pober JS, Cotran RS, Schlag G (1991) Expression of endothelial leukocyte adhesion molecule-1 (ELAM-1) in septic but not traumatic/hypovolemic shock in the baboon. Am J Pathol (in press)

Rice GE, Bevilacqua MP (1989) Tumor cell–endothelial interactions: an inducible endothelial cell surface molecule mediates melanoma cell adhesion. Science 246:1303–1306

Rice GE, Munro JM, Bevilacqua MP (1990) Inducible cell adhesion molecule 110 (INCAM-110) is an endothelial receptor for lymphocytes. J Exp Med 171:1369–1374

Rice GE, Munro JM, Corless C, Bevilacqua MP (1991) Vascular and nonvascular expression of INCAM-110. Am J Pathol 138:385–393

Ross R (1986) The pathogenesis of atherosclerosis: an update. N Engl J Med 314:488

Ross R, Glomset JA (1976) The pathogensis of atherosclerosis. N Engl J Med 295:369

Smith CW, Rothlein R, Hughes BJ et al. (1988) Recognition of an endothelial determinant for DC18-dependent human neutrophil adherence and transendothelial migration. J Clin Invest 82:1746–1756

Varani J, Ginsburg I, Schuger L et al. (1989) Endothelial cell killing by neutrophils. Am J Pathol 135:435–438

Walz G, Afuffo A, Kolanus W, Bevilacqua MP, Seed B (1990) Recognition of Elam-1 of the Sialyl-Le^x determinant on myeloid and tumor cells. Science 250:1132–1135

Ward PA, Varani J (1990) Mechanisms of neutrophil-mediated killing of endothelial cells (review). J Leukocyte Biol 48:97–102

Willms-Kretschmer K, Flax MH, Cotran RS (1967) The fine structure of the vascular response in hapten-specific delayed hypersensitivity and contact dermatitis. Lab Invest 17:334–349

Chapter 4

Basic Fibroblast Growth Factor in Vascular Development and Atherogenesis

W. Casscells

Introduction

Numerous laboratories are currently investigating the potential roles of polypeptide growth factors and oncogenes in atherosclerosis and related pathologies, such as restenosis after angioplasty and intimal proliferation in vessels of transplanted organs. Clearly, the rationale is that the development of these lesions involves the migration and proliferation of smooth muscle cells and monocyte–macrophages, and secretion of large amounts of extracellular proteoglycans and collagens (Ross et al. 1990a; Gerrity 1981; Gown et al. 1986; Wight 1989; Libby and Hansson 1991; Clowes and Schwartz 1985; McBride et al. 1988; Clowes et al. 1983; Manderson et al. 1989; Bulkley and Huchins 1977; Waller et al. 1984; Billingham 1989), well-known actions of growth factors, at least in vitro (Ross et al. 1990a; Gospodarowicz 1989; Baird and Böhlen 1990; Sporn and Roberts 1990). It is clear that most of the major classes of growth factors can by synthesized by activated macrophages (Ross et al. 1990a; Libby and Hansson 1991) and endothelial cells (Vlodavsky et al. 1987; Hannan et al. 1988; Speir et al. 1991; Baird and Ling 1987; Mansson et al. 1990; Gajdusek et al. 1980; Collins et al. 1987), and an increasing number of growth factors are being identified in vascular smooth muscle cells (Ross et al. 1990a; Libby and Hansson 1991; Baird and Ling 1987; Mansson et al. 1990; Weich et al. 1990; Gospodarowicz et al. 1988; Winkles et al. 1987; Cercek et al. 1990; Naftilan et al. 1989). Platelets have long been known to contain platelet-derived growth factors (PDGFs) (Ross et al. 1990a) and transforming growth factor-β (Sporn and Roberts 1990; Casscells et al. 1990a), and other mitogens such as serotonin (Corson et al. 1991).

Vascular Injury in Atherosclerosis

While the regulatory regions of many of the genes encoding these growth factors are now being characterized, there is still little information concerning the physiological regulation of these factors and their receptors in vivo. However, there is increasing evidence that a variety of growth factors may participate in wound healing, and there are suggestions that these effects may be complex and may vary with the particular type of wound (Grotendorst et al. 1985; Rappolee et al. 1988; Davidson et al. 1985; Buckley et al. 1985; Pierce et al. 1988; Mustoe et al. 1987). Thus, it may prove that a different pattern of growth factor expression will be detected in the acute and severe injury of angioplasty (McBride et al. 1988; Clowes et al. 1983; Manderson et al. 1989; Waller et al. 1984; Johnson et al. 1990; Dartsch et al. 1990; Clowes et al. 1989; Gravanis and Roubin 1989; Austin et al. 1985; Steele et al. 1985), compared to atherosclerosis (Ross et al. 1990a; Gown et al. 1986) where the injury is thought to be chronic, mild, and triggered not by a balloon but by oxidized lipids (Steinberg 1990), thrombus (Ross et al. 1990a; Schwartz et al. 1988; Gajdusek et al. 1986; Chesebro et al. 1982), inflammatory mediators (Libby and Hansson 1991; Gimbrone et al. 1990), viruses (Minick et al. 1979) and physical forces such as stretch and reduced shear (Ku et al. 1985; Ando et al. 1987; Davies et al. 1986; Ives and Mai 1991; McIntire et al. 1991; Langille et al. 1989).

The likelihood that the stimulus in atherosclerosis is mediated at least in part by endothelial cells subjected to subtle injury, without frank necrosis or desquamation, has generated intense interest in physiological factors that may influence the secretion of growth factors, and anti-growth factors, in endothelial cells (Wight 1989; Libby and Hansson 1991; Ives and Mai 1991; McIntire et al. 1991; Reidy and Schwartz 1983; Pober 1988; Munro and Cotran 1988; Fox and DiCorleto 1984).

There is considerable laboratory evidence that endothelial injury and loss predispose to smooth muscle cell migration and proliferation (Ross et al. 1990a; Wight 1989; Clowes et al. 1983; Tada and Reidy 1987; Stemerman et al. 1977). Loss of endothelial-derived heparin sulfates (Reilly et al. 1988; Clowes and Clowes 1989; Edelman et al. 1990; Castellot et al. 1973), and stimulation of synthesis and release of PDGFs (Ross et al. 1990a; Libby and Hansson 1991; Naftilan et al. 1989; Fox and DiCorlito 1984), interleukins (Libby and Hansson 1991) 1α, 1β and 6 and tumor necrosis factors (Libby and Hansson 1991), and perhaps endothelins, may be involved. There is also accumulating evidence that, in addition to an endothelial–smooth muscle cell paracrine mechanism, smooth muscle cells may utilize autocrine mechanisms of proliferation. Indeed the concept of atheroma as a benign tumor was suggested many years ago and is supported by (1) the focal distribution of lesions (Ross et al. 1990a; Ku et al. 1985), (2) certain clonal characteristics (Benditt and Benditt 1973), (3) the heterogeneity of smooth muscle cells (Manderson et al. 1989; Haudenschild and Grunwald 1985; Babaev et al. 1990), and (4) the discovery of transforming genes in smooth muscle cells in focus assays (Parkes et al. 1991).

Vascular Balloon Injury and Restenosis

Restenosis after angioplasty differs from atherosclerosis in a number of ways, including the severity of the injury, the speed of development (presumably due to a higher percentage of proliferating cells and shorter cell cycle) and in histological features. In analyzing specimens obtained by atherectomy from patients experiencing restenosis several months after percutaneous transluminal coronary angioplasty (PTCA), we (Flugelman et al. 1991), like others (Manderson et al. 1989; Waller et al. 1984; Johnson et al. 1990; Dartsch et al. 1990; Clowes et al. 1989; Gravanis and Roubin 1989; Austin et al. 1985), have noted that the majority of specimens reveal a fibrocellular process. Some regions are much more cellular than others. Occasional mitotic figures are seen. The vast majority of the cells have a mesenchymal appearance and stain deeply with phosphotungstic acid–hematoxylin and immunoreact with antibodies against α_1-smooth muscle actin. This strongly suggests that these cells are of smooth muscle cell lineage. By electron microscopy these cells have few bands of actin filaments but are rich in rough endoplasmic reticulum, consistent with the "synthetic phenotype" described in cultured smooth muscle cells and in smooth muscle cells proliferating after experimental balloon injury (Manderson et al. 1989; Owens 1989). In the atherectomy specimens we note few mononuclear cells but a moderate number of foam cells, many of which have a macrophage-like appearance and reveal lipid droplets when frozen sections are stained with oil Red-O. Thrombosis, hemorrhage and cholesterol clefts are uncommon, and calcification is detected very infrequently.

In contrast, atherectomy specimens from patients who have not previously undergone an angioplasty are generally less cellular, with denser fibrous tissue, a higher incidence of areas of necrosis, calcification and foam cells, and interstitial lipid. These differences suggest that the mechanism of restenosis after angioplasty may differ significantly from that of atherosclerosis. In keeping with this is the fact that several proven or potential anti-atherosclerotic regimens have generally been found to have little inhibitory effect on restenosis; these include smoking cessation, low cholesterol and fish oil diets or cholesterol-reducing drugs, and anti-inflammatory agents (Ip et al. 1990; Liu et al. 1989). To date, variable results have been reported with antihypertensive agents. Vasodilators may not only lower blood pressure, but may also chronically increase maximum lumen diameter by (1) increasing flow, (2) inhibiting cell proliferation by reducing intracellular (Ca^{2+}), in the case of calcium antagonists (Overturf 1990), or by blocking the mitogenic effects of norepinephrine (Blaes and Boissel 1983) and angiotensin II (Naftilan et al. 1989; Geisterfer et al. 1988; Bell and Madri 1989). Of these, the most promising is the blockade of angiotensin-converting enzyme, which was more successful than calcium antagonists in preventing neo-intimal proliferation in the balloon-injured rat carotid artery (Powell et al. 1989). Whether these findings can be replicated in humans will soon be known from recently concluded clinical trials.

However, the most intensively studied question has been the role of thrombosis, which is appealing since platelets contain growth factors such as PDGFs (Ross et al. 1990a) and TGF β_1 (Sporn and Roberts 1990; Bell and Madri 1989; Merwin et al. 1991; Owens et al. 1988; Goodman and Majack 1989;

Moses et al. 1990) and thrombi contain mitogenic substances such as fibronectin (Mosher 1989; Casscells et al. 1990b; Takasaki et al. 1990) and thrombin (Bar-Shavit et al. 1990). However, clinical studies suggest that antithrombotic regimens of various types have only been partially successful in preventing clinical restenosis (reviewed in Ip et al. 1990; Liu et al. 1989; Harker 1986). In animal models the induction of recurrent thrombosis can lead to thickening of the vessel wall and narrowing of the lumen particularly when unlysed thrombi become incorporated into the vessel wall (Steele et al. 1985; Schwartz et al. 1988). Nevertheless, after experimental balloon injury, antiplatelet and anticoagulant regimens (excluding heparin, which has direct inhibitory effects on smooth muscle growth) (Reilly et al. 1988; Clowes and Clowes 1989; Edelman et al. 1990; Castellot et al. 1973) have been only partially successful in preventing neo-intimal proliferation (Guyton and Karnovsky 1979; Fingerle et al. 1989; Clowes and Karnovsky 1977; Ingerman-Wojenski and Silver 1988; Willerson et al. 1989). One possible explanation for this might be that the acute thrombosis often observed in connection with balloon injury is generally transient (Clowes et al. 1983) and is in any case only one of several stimuli to smooth muscle proliferation. As previoulsy cited, other potential stimuli are: (1) stretch, which can stimulate proliferation of smooth muscle cells even without obvious cellular injury; (2) separation of cells (loss of contact inhibition); (3) infiltration by macrophages (which contain numerous growth factors); (4) release of growth factors stored by endothelial and smooth muscle cells; (5) loss of growth inhibitory substances contained in the vessel wall; and (6) the stimulation of synthesis of growth factors by the smooth muscle cells themselves.

Platelet-Derived Growth Factors

With regard to the last proposal – that of an autocrine or paracrine mechanism of smooth muscle migration and proliferation – the leading candidates are the platelet-derived growth factors (PDGFs) and their receptors (Ross et al. 1990a; Grotendorst et al. 1982; Terracio et al. 1988; La Rochelle et al. 1990). Although there are some discrepancies in these studies, there is increasing evidence that PDGFs are expressed at higher levels in proliferating smooth muscle cells in experimentally injured, atherosclerotic and transplanted vessels than in normal vessels (Wilcox et al. 1988; Barrett and Benditt 1988; Libby et al. 1988; Rubin et al. 1988; Ross et al. 1990b). The participation of other growth factors has not been excluded, but may simply not be necessary given the extensive evidence that smooth muscle cells, endothelial cells and macrophages all can synthesize and release one or more members of the PDGF family, and that these PDGFs are, to varying degrees, chemotactic and mitogenic for smooth muscle cells, and chemotactic for monocytes. Moreover, a number of other mitogens are now thought to work in large part indirectly, by stimulating expression of PDGFs (Libby and Hansson 1991; Baird and Böhlen 1990; Naftilan et al. 1989). Thus there was no compelling reason to suspect the involvement of basic fibroblast growth factor (Gospodarowicz 1989; Baird and Böhlen 1990; Rifkin and Moscatelli 1989), though several of its actions in vitro suggested it could play a role in atherogenesis (Gospodarowicz et al. 1981; Hoshi et al. 1988; Chen et al. 1988; Klagsbrun and Edelman 1989; Casscells et al. 1990c).

Fibroblast Growth Factors

Basic fibroblast growth factor (bFGF), an 18-kDa heparin-binding peptide, is the best-understood member of an eight-gene FGF family (Burgess and Maciag 1989). The name bFGF is a misnomer, since bFGF is found in many types of cells, besides fibroblasts, and is mitogenic for most cells of neuroectodermal, mesodermal, and even endodermal origin (Gospodarowicz 1989; Baird and Böhlen 1990; Rifkin and Moscatelli 1989). Several groups have reported mitogenic effects in smooth muscle cells but two groups have found no such effects, suggesting the possibility that the mitogenic and chemotactic effects may depend on the assay conditions (Risau 1986; Grotendorst et al. 1982). Alternatively, these negative reports may indicate the existence of subpopulations of smooth muscle cells that lack FGF receptors or have been activated by mutations acting distal to bFGF in the cell cycle.

Other reported properties of bFGF include angiogenesis, induction of embryonic mesoderm, regulation of certain endocrine functions, and of synthesis and degradation of various components of the extracellular matrix, enhancement of cell survival, and stimulation of neuronal differentiation (Gospodarowicz 1989; Baird and Böhlen 1990; Rifkin and Moscatelli 1989). The extent to which these properties are utilized in vivo has yet to be determined. The addition of bFGF to bioassays, however, has been shown to promote angiogenesis (Folkman and Klagsbrun 1987) and wound healing (McGee et al. 1988; Greenhalgh et al. 1990).

Expression of bFGF in Vascular Development and Injury

Three recent findings from this laboratory led us to ask whether bFGF might be involved in the smooth muscle proliferation following vascular injury. The first was the finding that smooth muscle cells in the normal adult rat aorta contain only a small amount of bFGF compared to an aliquot of the same cells proliferating in culture. Moreover, the cultured cells bound severalfold more radiolabeled bFGF and expressed readily detectable amounts of bFGF mRNA, whereas the freshly isolated cells, or the cells in the intact artery, did not (Speir et al. 1991).

The second relevant finding was that infusion of acidic FGF (the product of a separate gene with similar mitogenic targets in vitro (Burgess and Maciag 1989) into ischemic myocardial tissue resulted in smooth muscle cell proliferation in the arterioles (Banai et al. 1991).

Third, we found more bFGF immunoreactivity in the medial layers of embryonic rat coronary arteries than in the adult rat medial layer (Spirito et al. 1991).

To the extent that cells in culture are "wounded", and to the extent that wounding recapitulates some features of embryonic development, it thus seemed likely that balloon injury of the rat carotid artery might lead to a

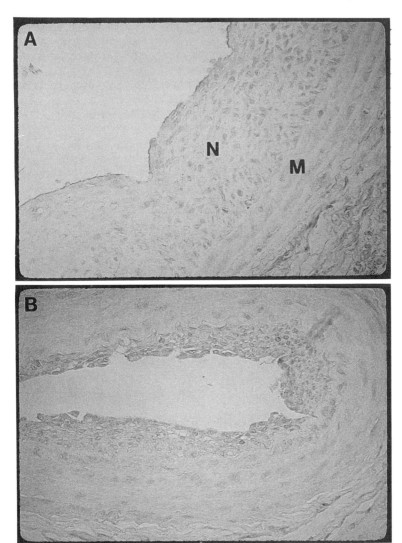

Fig. 4.1. Expression of bFGF in proliferating vascular smooth muscle. **A:** Increase in bFGF immunoreactivity in carotid medial (smooth muscle) layer of adult rat ten days after balloon injury. By ten days the neo-intima (N) is denuded of endothelial cells and is larger than the medial layer (M). The brown (immunoperoxidase-positive) cells that are immunoreactive for bFGF are prominent in the proliferating smooth muscle cells (SMC) of the neo-intima and rare in medial layer, which is relatively quiescent. After an initial loss of bFGF immunoreactivity, an increase in bFGF immunoreactivity in medial SMC was first noted by 24–48 hours, coinciding with the initiation of DNA synthesis and preceding the onset of SMC migration. Formalin-fixed, 6-μm paraffin sections of 12-week-old male rat carotids were etched with hyaluronidase, blocked and stained by either the indirect alkaline phosphatase or ABC method, as described by Casscells et al. (1990c). Identical results were obtained with A. Baird's N-terminally directed polyclonal anti-bFGF IgG "773" (A) and by anti-bFGF monoclonal No. 78 (B), a gift from Dr Y. Oka (Takeda Corp, Japan). Almost no stain was noted when 1° antisera were pre-absorbed with tenfold excess bFGF or replaced by normal rabbit serum (**B**) and little immunoreactivity developed if sections were pre-washed in 2 m NaCl to remove heparin-bound bFGF. The 1° antisera have 0%–1% cross-reactivity with aFGF in Western blotting, and the N-terminus of bFGF (against which 773 and 78 are directed) has no significant homology to other members of the FGF family.

re-expression of bFGF. The first indication that this might indeed be the case came from immunohistological studies. Normal rat carotid and aortic vessels had little bFGF immunoreactivity in their medial (smooth muscle) layers. Balloon injury resulted in a variable loss of immunoreactivity, probably indicating release of bFGF, followed by an increase in intracellular bFGF immunoreactivity from 24 hours to ten days (Fig. 4.1). Our confidence in these studies was increased when a variety of antisera directed against different epitopes of bFGF yielded the same result. Moreover, with several of these antisera bFGF was the only or the predominant band in immunoblots of crude vessel lysates. These bands, like the immunostaining of the tissues, could not be detected if the antisera were pre-adsorbed with the peptide immunogens, or with recombinant bFGF (Casscells 1991). Recently we have obtained similar immunohistological results in atherectomy specimens from human coronary arteries (Flugelman et al. 1991).

The likelihood that the increased bFGF expression serves a mitogenic function (as opposed to one of bFGF's non-mitogenic functions) was indicated not only by bFGF's mitogenic effect on cultured vascular smooth muscle cells (described below), but also by the in vivo mitogenic effect of aFGF described above (Banai et al. 1991). However, these data do not guarantee that neutralizing antisera to bFGF will inhibit the smooth muscle proliferation in experimental vascular injury.

Prospects for Anti-bFGFs

Our first concern was based on the fact that numerous growth factors can promote proliferation of smooth muscle cells, and it is not yet known whether each of these factors is an absolute requirement for proliferation, perhaps in a specific sequence in the cell cycle, or whether the growth factors act in an additive fashion, or whether they are redundant, such that any one growth factor is sufficient. Our second concern was based on our recent finding that bFGF is localized not only in smooth muscle cell cytopolasm and extracellular matrix, but also is associated with nuclear chromatin and nucleoli, particularly in actively growing cells. If nuclear localization is important for mitogenic activity (as suggested for aFGF) and if there is direct translocation of bFGF to the nucleus (i.e., without leaving the cell and binding to receptors on the cell surface), then neutralizing antibodies would be ineffective in preventing proliferation, since their entry into most cells is very inefficient.

A third problem with antibody therapies is the high likelihood of developing neutralizing antibodies against the anti-bFGF antibodies.

Because of the difficulty and expense of addressing these questions with injections of neutralizing anti-bFGF antibodies in vivo, we first chose to model the acute vascular injury by studying the proliferation of smooth muscle cells in vitro. We used high concentrations of serum to model the platelet aggregation, release and coagulation of acute balloon injury, and used very low serum concentrations to model the non-proliferating, uninjured vessel.

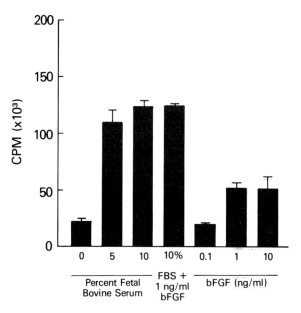

Fig. 4.2. Greater mitogenic effect of bFGF in 1% than in 10% fetal bovine serum (FBS). Smooth muscle cells (SMC) were isolated by enzyme digestion of medial layers from 7-week-old male Sprague–Dawley rats, grown in M199/10% FBS without bFGF, and split weekly at a 1/4 ratio. Confluent cells at passage 10, displaying the characteristic multilayered hill and valley appearance, phase-dense cytoplasm, and reactivity with A. Gown's monoclonal anti-smooth muscle α-actin antibody were exposed to serum-free M199 (without bFGF) for 72 hours, at which time FBS (Hyclone) and/or bFGF (Drs P. Barr and L. Cousens, Chiron) plus insulin 6.25 μg/ml, transferrin 6.25 μg/ml, selenium 6.25 ng/ml and 1 μCi/well 3-HTdr (6.7 Ci/mmol, American Radiolabeled Chemicals) were added, followed by scintillation counting 36 hours later, as described. The experiment was repeated twice. Similar results were obtained when cells were counted by hemocytometer after 72 hours exposure to additives. Bars indicate means ± SD of quintuplicate wells. In P6 cells from another isolation (not shown) we noted up to 50% increase in final SMC density. However, adding bFGF was generally found not to enhance proliferation of subconfluent SMC in optimal serum.

As expected, in 0.5% serum the smooth muscle cells proliferated slowly, and were stimulated by the addition of either bFGF or serum. However, in saturating concentrations of serum (which contains no detectable bFGF), the addition of bFGF to the culture medium provided little or no additional mitogenic stimulation (Fig. 4.2). Moreover, the addition of saturating concentrations of neutralizing antibodies to bFGF, which caused a significant inhibition of the added mitogenic effect of exogenous bFGF, had little effect on smooth muscle cell proliferation in either 0.5% or 10% serum (Fig. 4.3) (Casscells 1991). These data suggested that in optimal serum concentrations smooth muscle cells are proliferating under the influence of the multiple growth factors present in serum, in addition to the use of growth factors synthesized by the smooth muscle cells themselves, and that the addition of bFGF was essentially redundant.

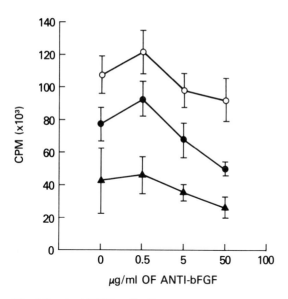

Fig. 4.3. Anti-bFGF antibodies neutralize the mitogenic effect of added bFGF but have little effect in serum alone. Rat aortic cells, P_{12}, were plated at $30\,000/cm^2$ in 1% (bottom curve) or 10% (top curve) FBS in M199 and 24 hours later given 0–$50\,\mu g/ml$ anti-bFGF "dog", a highly specific polyclonal antibody against whole human recombinant bFGF, kindly donated by D. Gospodarowicz. Cells were exposed 12 hours later to ^{3}HTdR ($1\,\mu Ci/200\,\mu l$ well) and bound ^{3}H was counted 24 hours after that. The experiment was repeated once, and twice more using P_7 rabbit SMC, with similar results. Monoclonal anti-bFGF's DG-2[58] (a gift of T. Reilly, DuPont, Wilmington, Delaware) and Fm-1 (UBI, Lake Placid, New York) gave similar results at tenfold higher concentrations.

Targeted Toxin Therapy

Although in vivo responses may differ from in vitro models, these data suggested that neutralizing antibodies to bFGF might have only a modest effect on smooth muscle proliferation in vivo. This led us to try an alternative approach – that of using bFGF to carry a ribosome-inactivating toxin into those proliferating smooth muscle cells which express FGF receptors. Although similar approaches (toxins linked to antibodies against specific cell surface markers) have had only limited success in the treatment of cancer (Vitetta and Thorpe 1991) we were encouraged by the fact that bFGF is a highly internalized ligand, unlike many cell surface antigens. Moreover, it has previously been shown that bFGF is degraded only very slowly after internalization (Lappi et al. 1989), thus raising the possibility that the enzyme might remain free in the cytosol long enough to act. Finally, we reasoned that smooth muscle cells would be less able than cancer cells to evade the toxin by rapidly generating cells that lack FGF receptors. In contrast, in many tumors the antigens on the cell surface change rapidly. Thus, as many as 99.99% of the malignant cells must be killed to have a clinically significant impact, whereas a 90% loss of smooth muscle cells might suffice since the smooth muscle accumulation, however excessive, eventually ceases.

The toxin we selected, saporin, is an *N*-glycosidase derived from the seed of the plant *Saponaria officinalis*. Saporin is taken up slowly and probably non-specifically by animal cells. However, once inside a cell, saporin inactivates the 28S rRNA of the 60S ribosomal subunit by clearing adenine from ribose. Saporin was conjugated to bFGF by a hetero-bifunctional cross-linking reagent by Douglas Lappi and Andrew Baird, purified on heparin–Sepharose and generously made available for these studies (Lappi et al. 1989).

In sparse vascular smooth muscle cells proliferating in 10% serum, bFGF–saporin inhibited protein synthesis within 14 hours at a concentration of 10 nM and by 48 hours at a concentration of 0.1 nM. In contrast, saporin alone inhibited protein synthesis only at 500-fold higher concentrations. DNA synthesis was also inhibited by bFGF–saporin and, as with protein synthesis, the effect was prevented if the bFGF–saporin conjugate was incubated with an excess of bFGF, confirming that the conjugate toxin acted via the FGF receptor. From 48 to 96 hours the conjugate produced smooth muscle cell killing. Very low doses prevented an increase in smooth muscle cell number, apparently without cell killing.

These effects were best seen in subconfluent cells proliferating in the presence of 10% serum. Cell killing required 1000-fold higher concentrations in dense, serum-starved cells, perhaps reflecting lower requirements for protein synthesis or an absence of FGF receptors (Fig. 4.4). The relative immunity of quiescent (non-proliferating) cells from the toxic effect of bFGF–saporin suggested that non-proliferating cells in vivo might also be safe, whereas proliferating smooth muscle cells would selectively be killed or inhibited. Indeed, preliminary data from Scatchard, Western and Northern assays indicate that few FGF receptors

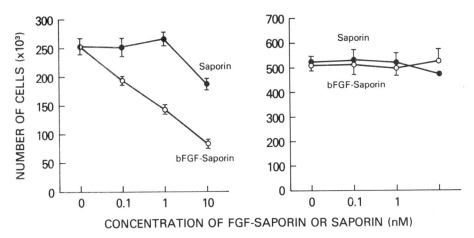

Fig. 4.4. Quiescent smooth muscle cells are not killed by bFGF–saporin. Left panel: subconfluent $(2\times10^4/cm^2)$ P_{12} rat aortic SMC in 10% FBS were exposed to indicated concentrations of saporin or bFGF–saporin for 1 hour, then washed for 10 minutes with 10 µg/ml heparin sulfate to remove matrix-bound bFGF–saporin, refed 10% FBS/M199 and trypsinized and counted 96 hours later. Right panel: confluent $(2\times10^5/cm^2)$ cells were made more quiescent by 72 hours in serum-free M199, then exposed to bFGF–saporin (open circles) or saporin (closed circles), washed, refed, and counted as above. Bars indicate standard deviations of triplicate wells counted by hemocytometer. The experiment was repeated once with similar results.

are expressed in normal vessels, whereas many are expressed in proliferating smooth muscle cells after balloon injury. Consequently, studies are currently under way to determine if intravenous bFGF–saporin can be used as a self-targeting and selective inhibitor of smooth muscle proliferation in vivo. The possibility that radiolabeled bFGF may also serve as an imaging agent to detect FGF receptor expression in the vessel wall after injury, and perhaps predict restenosis, also appears to deserve investigation. Equally intriguing is the possibility that bFGF and FGF receptor expression in endothelial and smooth muscle cells are involved in the pathogenesis of atherosclerosis.

References

Ando J, Nomura H, Kamiya S (1987) The effect of fluid shear stress on the migration and proliferation of cultured smooth muscle cell. Microvasc Res 33:62–70

Austin GE, Ratliff NB, Hollman J, Tabei S, Phillips DF (1985) Intimal proliferation of smooth muscle cells as an explanation for recurrent coronary artery stenosis after percutaneous transluminal coronary angioplasty. J Am Coll Cardiol 6:369–375

Babaev VR, Bobryshev YV, Stenina OV, Tararak EM, Gabbiani G (1990) Heterogeneity of smooth muscle cells in atheromatous plaque of human aorta. Am J Pathol 136:1031–1042

Baird A, Böhlen P (1990) Fibroblast growth factors. In: Sporn MB, Roberts AB (eds) Handbook of experimental pharmacology. Peptide growth factors and their receptors. Springer-Verlag, Berlin, pp 369–418

Baird A, Ling N (1987) Fibroblast growth factors are present in the extracellular matrix produced by endothelial cells in vitro: implications for a role of heparinase-like enzymes in the neovascular response. Biochem Biophys Res Commun 142:428–435

Banai S, Jaklitsch M, Casscells W, Shou M, Shrivastav S, Correa R, Epstein SE, Unger EF (1991) Effects of acidic fibroblast growth factor on normal and ischemic myocardium. Circ Res (accepted for publication)

Barrett TB, Benditt EP (1988) Platelet-derived growth factor gene expression in human atherosclerotic plaque and normal artery wall. Proc Natl Acad Sci USA 85:2810–2814

Bar-Shavit R, Benezra M, Eldort A, Hy-Am H, Fenton JW II, Wilners GD, Vlodavsky I (1990) Thrombin immobilized to extracellular matrix is a potent mitogen for vascular smooth muscle cells: nonenzymatic mode of action. Cell Regulation 1:453–463

Bell L, Madri JA (1989) Effect of platelet factors on migration of cultured bovine aortic endothelial and smooth muscle cells. Circ Res 65:1057–1065

Benditt EP, Benditt JM (1973) Evidence for a monoclonal origin of human atherosclerosis plaques. Proc Natl Acad Sci USA 70:1753–1756

Billingham ME (1989) Graft coronary disease: the lesion and the patient. Transplant Proc 21:3665–3666

Blaes N, Boissel JP (1983) Growth-stimulating effect of catecholamines on rat aortic smooth muscle cells in culture. J Cell Physiol 116:167–175

Buckley A, Davidson JM, Kameroth CD, Wolt TB, Woodward SC (1985) Sustained release of epidermal growth factor accelerates wound repair. Proc Natl Acad Sci USA 82:7340–7344

Bulkley BH, Hutchins GM (1977) Accelerated "atherosclerosis": a morphological study of 97 saphenous vein coronary artery bypass grafts. Circulation 50:163–169

Burgess WH, Maciag T (1989) The heparin-binding (fibroblast) growth factor family of proteins. Annu Rev Biochem 58:575–606

Casscells W (1991) The bFGF system is involved in vascular development and can be manipulated to inhibit smooth muscle proliferation. J Cell Biochem 15C:98

Casscells W, Bazoberry F, Speir E, Thompson N, Flanders K, Kondaiah P, Ferrans V, Epstein SE, Sporn M (1990a) Transforming growth factor B1 in normal heart and in myocardial infarction. Ann NY Acad Sci 593:148–160

Casscells W, Kimura H, Sanchez JA, Yu Z-X, Ferrans VJ (1990b) Immunohistochemical study of fibronectin in experimental myocardial infarction. Am J Pathol 137:801–810

Casscells W, Speir E, Sasse J, Klagsbrun M, Allen P, Lee M, Calvo B, Chiba M, Haggroth L, Folkman J, Epstein SE (1990c) Isolation, characterization and localization of heparin-binding growth factors in the heart. J Clin Invest 85:433–441

Castellot J, Cochran D, Karnovsky M (1973) Effect of heparin on vascular smooth muscle cells. I. Cell metabolism. J Cell Physiol 124:21–38

Cercek B, Fishbein MC, Forrester JS, Helfant RH, Fagin JA (1990) Induction of insulin-like growth factor 2 messenger RNA in rat aorta after balloon denudation. Circ Res 66:1755–1760

Chen J-K, Hoshi H, McKeehan WL (1988) Heparin-binding growth factor type one and platelet-derived growth factor are required from the optimal expression of cell surface low density lipoprotein receptor binding activity in human adult arterial smooth muscle cells. In Vitro Cell Dev Biol 24:199–204

Chesebro JH, Clements LP, Fuster V, Elveback LR, Smith HC et al. (1982) A platelet-inhibitor drug trial in coronary-artery bypass operations: benefit of perioperative dipyridamole and aspirin therapy on early postoperative vein-graft patency. N Engl J Med 307:73–78

Clowes AW, Clowes MM (1989) Inhibition of smooth muscle cell proliferation by heparin molecules. Transplant Proc 21:3700–3701

Clowes AW, Karnovsky MJ (1977) Failure of certian antiplatelet drugs to affect myointimal thickening following arterial endothelial injury in the rat. Lab Invest 36:452–464

Clowes AW, Schwartz SM (1985) Significance of quiescent smooth muscle migration in the injured rat carotid artery. Circ Res 56:139–145

Clowes AW, Reidy MA, Clowes MM (1983) Kinetics of cellular proliferation after arterial injury. I. Smooth muscle growth in the absence of endothelium. Lab Invest 49:327–333

Clowes AW, Clowes MM, Fingerle J, Reidy MA (1989) Kinetics of cellular proliferation after arterial injury. V. Role of acute distension in the induction of smooth muscle proliferation. Lab Invest 60:360–364

Collins T, Pober JS, Gimbrone MA Jr, Betsholtz C, Westermark B, Heldin CH (1987) Cultured human endothelial cells express platelet-derived growth factor A chain. Am J Pathol 126:7–12

Corson M, Alexander RW, Berk BC (1991) THe phospholipase C coupled serotonin receptor mediating rat aortic smooth muscle cell growth is the 5-HT$_2$ sub-type. J Cell Biochem 15C:123

Dartsch PC, Voisard R, Bauriedel G, Hofling B, Betz E (1990) Growth characteristics and cytoskeletal organization of cultured smooth muscle cells from human primary stenosing and restenosing lesions. Arteriosclerosis 10: 62–75

Davidson JM, Klagsbrun M, Hill KE, Buckley A, Sullivan R, Brewer PS, Woodward SC (1985) Accelerated wound repair, cell proliferation and collagen accumulation are produced by a cartilage-derived growth factor. J Cell Biol 100:1219–1227

Davies PF, Kemuzzi A, Gordon EJ, Dewey CF Jr, Gimbrone MA Jr (1986) Turbulent fluid shear stress induces vascular endothelial cell turnover in vitro. Proc Natl Acad Sci USA 83:2114–2117

Edelman ER, Adams DH, Karnovsky MJ (1990) Effect of controlled adventitial heparin delivery on smooth muscle cell proliferation following endothelial injury. Proc Natl Acad Sci USA 87:3773–3777

Fingerle J, Johnson R, Clowes AW, Majesky MW, Reidy MA (1989) Role of platelets in smooth muscle cell proliferation and migration after vascular injury in rat carotid artery. Proc Natl Acad Sci USA 86:8412–8416

Flugelman MY, Correa R, Yu Z-X, Keren G, Leon MB, Satler LF, Kent KM, Casscells W, Epstein SE (1991) Fibroblast growth factors are expressed in coronary lesions of patients with unstable angina pectoris and those who have post-angioplasty restenosis. J Am Coll Cardiol 17:73A

Folkman J, Klagsbrun M (1987) Angiogenic factors. Science 235:442–447

Fox PL, DiCorleto PE (1984) Regulation of production of a platelet-derived growth factor-like protein by cultured bovine aortic endothelial cells. J Cell Physiol 121:298–308

Gajdusek C, DiCorleto PE, Ross R, Schwartz SM (1980) An endothelial cell-derived growth factor. J Cell Biol 85:467–472

Gajdusek C, Carbon S, Ross R, Nawroth P, Stern D (1986) Activation of coagulation releases endothelial cell mitogens. J Cell Biol 103:419–428

Geisterfer AA, Peach MJ, Owens GK (1988) Angiotensin II induces hypertrophy, not hyperplasia, of cultured rat aortic smooth muscle cells. Circ Res 62:749–756

Gerrity RG (1981) The role of the monocyte in atherogenesis. 1. Transition of blood-borne monocytes into foam cells in fatty lesions. Am J Pathol 103:181–190

Gimbrone MA Jr, Bevilacqua MP, Cybulsky MI (1990) Endothelial-dependent mechanisms of leukocyte adhesion in inflammation and atherosclerosis. Ann NY Acad Sci 598:77–85

Goodman LV, Majack RA (1989) Vascular smooth muscle cells express distinct transforming growth factor-β receptor phenotypes as a function of cell density in culture. J Biol Chem 264:5241–5244

Gospodarowicz D (1989) Fibroblast growth factor. Crit Rev Oncogen 1:1–26

Gospodarowicz D, Hirabayashi K, Giguere L, Tauber JP (1981) Factors controlling the proliferative rate, final cell density and life span of bovine vascular smooth muscle cells in culture. J Cell Biol 89:568–578

Gospodarowicz D, Ferrara N, Haaparanta T, Neufeld G (1988) Basic fibroblast growth factor: expression in cultured bovine vascular smooth muscle cells. Eur J Cell Biol 46:144–151

Gown AM, Tsukada T, Ross R (1986) Human atherosclerosis: immunocytochemical analysis of the cellular composition of human atherosclerotic lesions. Am J Pathol 125:191–207

Gravanis MB, Roubin GS (1989) Histopathologic phenomena at the site of percutaneous transluminal coronary angioplasty: the problem of restenosis. Hum Pathol 20:477–485

Greenhalgh DG, Sprugel KH, Murray MJ, Ross R (1990) PDGF and FGF stimulate wound healing in the genetically diabetic mouse. Am J Pathol 136:1235–1246

Grotendorst GR, Chang T, Seppa HEJ, Kleinman HK, Martin GR (1982) Platelet-derived growth factor is a chemoattractant for vascular smooth muscle cells. J Cell Physiol 113:261–266

Grotendorst GR, Martin GR, Pancev D, Sodek J, Harvey AK (1985) Stimulation of granulation tissue formation by platelet-derived growth factor in normal and diabetic rats. J Clin Invest 76:2323–2329

Guyton JR, Karnovsky MJ (1979) Smooth muscle cell proliferation in the occluded rat carotid artery: lack of requirement for luminal platelets. Am J Pathol 94:585–602

Hannan RL, Kourembana S, Flanders KC, Rogel S, Roberts AB, Faller DV, Klagsbrun M (1988) Endothelial cells synthesize basic fibroblast growth factor and transforming growth factor beta. Growth Factors 1:1–17

Harker LA (1986) Clinical trials evaluating platelet-modifying drugs in patients with atherosclerotic cardiovascular disease and thrombosis. Circulation 2:206–223

Haudenschild CC, Grunwald J (1985) Proliferative heterogeneity of vascular smooth muscle cells and its alteration by injury. Exp Cell Res 157:364–370

Hoshi H, Kan M, Chen J-K, McKeehan W (1988) Comparative endocrinology–paracrinology–autocrinology of human adult large vessel endothelial and smooth muscle cells. In Vitro Cell Dev Biol 24:309–320

Ingerman-Wojenski CM, Silver MJ (1988) Model system to study interaction of platelets with damaged arterial wall. II. Inhibition of smooth muscle cell proliferation by dipyridamole and AH-P719. Exp Mol Path 48:116–134

Ip JH, Fuster V, Badimon L, Badimon J, Taubman MB, Chesebro JH (1990) Syndromes of accelerated atherosclerosis: role of vascular injury and smooth muscle cell proliferation. J Am Coll Cardiol 15:1667–1687

Ives HE, Mai Q (1991) Effects of cyclic stretch on growth, ion transport and gene expression in vascular smooth muscle cells. J Cell Biochem 15C:124

Johnson DE, Hinohara T, Selmon MR, Braden LJ, Simpson JB (1990) Primary peripheral arterial stenosis and restenosis excised by transluminal atherectomy: a histopathologic study. J Am Coll Cardiol 15:419–425

Klagsbrun M, Edelman ER (1989) Biological and biochemical properties of fibroblast growth factor: implications for the pathogenesis of atherosclerosis. Atherosclerosis 9:269–278

Ku DN, Giddens DP, Sarins CK, Glagov S (1985) Pulsatile flow and atherosclerosis in human carotid bifurcation: positive correlation between plaque location and low and oscillatory shear stress. Arteriosclerosis 5:292–302

Langille BL, Bendeck MP, Kelley FW (1989) Adaptations of carotid arteries of young and mature rabbit to reduced carotid blood flow. Am J Physiol 256:H931–H939

Lappi DA, Martineau D, Baird A (1989) Biological and chemical characterization of basic FGF–saporin mitotoxin. Biochem Biophys Res Commun 160:917–923

La Rochelle WJ, Giese N, May-Siroff M, Robbins KC, Aaronson SA (1990) Molecular localization of the transforming and secretory properties of PDGF A and PDGF B. Science 248:1541–1544

Libby P, Hansson GK (1991) Involvement of the immune system in human atherogenesis. Lab Invest 64:5–15

Libby P, Warner SJC, Salomon RN, Birinyi LK (1988) Production of platelet derived growth factor-like mitogen by smooth muscle cells from human atheroma. N Engl J Med 318:1493–1498

Liu MW, Roubin GS, King SB (1989) Restenosis after coronary angioplasty: potential biologic determinants and role of intimal hyperplasia. Circulation 79:1374–1378

Manderson JA, Mosse PR, Safatrom JA, Young SB, Campbell GR (1989) Balloon catheter injury to rabbit carotid artery. I. Changes in smooth muscle phenotype. Arteriosclerosis 9:289–298

Mansson PE, Marlak M, Sawada H, Kan M, McKeehan WL (1990) Heparin-binding (fibroblast) growth factors type one and two genes are co-expressed in proliferating normal human vascular endothelial and smooth muscle cells in culture. In Vitro Cell Dev Biol 26:209–212

McBride W, Lange RA, Hillis DL (1988) Restenosis after successful coronary angioplasty: pathophysiology and prevention. N Engl J Med 318:1734–1737

McGee GS, Davidson JM, Buckley A, Sommer A, Woodward SC, Aquino AM, Barbour R, Demetriou AA (1988) Recombinant basic fibroblast growth factor accelerates wound healing. J Surg Res 45:145–153

McIntire LV, Hamond SL, Sharefkin JB, Eskin SG (1991) Regulation of gene expression in endothelial cells exposed to laminar shear stress. J Cell Biochem 15C:136

Merwin JR, Newman W, Beale LD, Tucker A, Madri J (1991) Vascular cells respond differently to transforming growth factors beta 1 and beta 2 in vitro. Am J Pathol 138:37–51

Minick CR, Fabricant CG, Fabricant J, Litrenta MM (1979) Atheroarteriosclerosis induced by infection with a herpes virus. Am J Pathol 96:673–684

Moses HL, Yang EY, Pietenpol JA (1990) TGF-β stimulation and inhibition of cell proliferation: new mechanistic insights. Cell 63:245–247

Mosher DF (ed) (1989) Fibronectin. Academic Press, San Diego

Munro JM, Cotran RS (1988) The pathogenesis of atherosclerosis: atherogenesis and inflammation. Lab Invest 58:249–261

Mustoe TA, Pierce GF, Thomason A, Gramates P, Sporn MB, Deuel TF (1987) Accelerated healing of incisional wounds in rats induced by transforming growth factor-β. Science (Wash DC) 234:1333–1335

Naftilan AJ, Pratt RE, Dzau VJ (1989) Induction of platelet-derived growth factor A-chain and c-myc gene expressions by angiotensin II in cultured rat vascular smooth muscle cells. J Clin Invest 83:1419–1424

Overturf M (1990) Are calcium ion antagonists effective anti-atherogenic agents? Arteriosclerosis 10:961–962

Owens GK (1989) Control of hypertrophic versus hyperplastic growth of vascular smooth muscle cells. Am J Physiol 252:H1755–H1765

Owens GK, Geisterfer AAT, Yang Y W-H Komoriya A (1988) TGF-β-induced growth inhibition and cellular hypertrophy in cultured vascular smooth muscle cells. J Cell Biol 107:771–780

Parkes JL, Cardell RR, Hubbard FC Jr, Meltzer A, Penn A (1991) Cultured human atherosclerotic plaque smooth muscle cells retain transforming potential and display enhanced expression of the myc proto-oncogene. Am J Pathol 138:765–775

Pierce GF, Mustoe TA, Senior RM, Reed J, Griffin GL, Thomason A, Deuel TF (1988) In vivo incisional wound healing augmented by platelet-derived growth factors and recombinant c-sis gene homodimeric proteins. J Exp Med 167:974–987

Pober JS (1988) Cytokine-mediated activation of vascular endothelium: physiology and pathology. Am J Pathol 133:425–433

Powell JS, Clozel JP, Mullër RKM, Kuhn H, Hefti F, Hosang M, Baumgartner HR (1989) Inhibitors of angiotensin-converting enzyme prevent myointimal proliferation after vascular injury. Science 245:186–188

Rappolee DA, Mark D, Banda MJ, Werb Z (1988) Wound macrophages express TGF-α and other growth factors in vivo: analysis by mRNA phenotyping. Science (Wash DC) 241:708–712

Reidy MA, Schwartz SM (1983) Endothelial injury and regeneration. IV. Endotoxin: a nondenuding injury to aortic endothelium. Lab Invest 48:25–34

Reilly C, Fritze LMS, Rosenberg RD (1988) Heparin-like molecules regulate the number of epidermal growth factor receptors on vascular smooth muscle cells. J Cell Physiol 136:23–32

Rifkin DB, Moscatelli D (1989) Recent developments in the cell biology of basic fibroblast growth factor. J Cell Biol 109:1–6

Risau W (1986) Developing brain produces an angiogenic factor. Proc Natl Acad Sci USA 83:3855–3859

Ross R, Masuda J, Raines EW (1990a) Cellular interactions, growth factors, and smooth muscle proliferation in atherogenesis. Ann NY Acad Sci 598:102–111

Ross R, Masuda J, Raines EW, Gown AM, Katsuda S, Sasahara M, Malden LT, Masuko H, Sato H (1990b) Localization of PDGF-B protein in macrophages in all phases of atherogenesis. Science 248:1009–1012

Rubin L, Tingstrom A, Hansson GK, Larsson E, Ronnstrand L, Klareskog L, Claesson-Welsh L, Heldon C-H, Fellstrom B, Terracio L (1988) Induction of B-type receptors for platelet-derived growth factor in vascular inflammation: possible implications for development of vascular proliferative lesions. Lancet i:1353–1356

Schwartz CJ, Valente AJ, Kelly JL, Sprague EA, Edwards EH (1988) Thrombosis and the development of atherosclerosis: Rokitansky revisited. Semin Thromb Hemost 14:189–194

Speir E, Sasse J, Shrivastav S, Casscells W (1991) Culture-induced increase in acidic and basic fibroblast growth factor activities and their association with the nuclei of vascular endothelial and smooth muscle cells. J Cell Physiol 147:362–373

Spirito P, Fu Y-M, Yu Z-X, Epstein SE, Casscells W (1991) Immunohistochemical localization of basic and acidic fibroblast growth factors in the developing rat heart. Circulation (in press)

Sporn MB, Roberts AB (1990) TGF-β: problems and prospects. Cell Regul 1:875–882

Steele PM, Chesebro JH, Stanson AW, Holmes DR Jr, Dewanjee MK, Badimon L, Fuster F (1985) Balloon angioplasty: natural history of the pathophysiological response to injury in a pig model. Circ Res 57:105–112

Steinberg D (1990) Arterial metabolism of lipoproteins in relation to atherogenesis. Ann NY Acad Sci 598:125–135

Stemerman MB, Spaet TH, Pitlick F, Cintron J, Lejnieks I, Tiell ML (1977) Intimal healing: the pattern of reendothelialization and intimal thickening. Am J Pathol 87:125–142

Tada T, Reidy MA (1987) Endothelial regneration. IX. Arterial injury followed by rapid endothelial repair induces smooth-muscle-cell proliferation but not intimal thickening. Am J Pathol 129:429–433

Takasaki I, Chobanian AV, Sarzani R, Brecher P (1990) Effect of hypertension on fibronectin expression in the rat aorta. J Biol Chem 265:21935–21939

Terracio L, Ronnstrand L, Tingstrom A, Rubin K, Claesson-Welsh L, Funa K, Heldin C-H (1988) Induction of platelet-derived growth factor receptor expression in smooth muscle cells and fibroblasts upon tissue culturing. J Cell Biol 107:1947–1957

Vitetta E, Thorpe PE (1991) Immunotoxins. In: DeVita V, Hellman S, Rosenberg S (eds) Biologic therapy of cancer: principles and practice. Lippincott, Philadelphia (in press)

Vlodavsky I, Fridman R, Sullivan R, Sasse J, Klagsbrun M (1987) Aortic endothelial cells synthesize basic fibroblast growth factor which remains cell associated and platelet-derived growth factor-like protein which is secreted. J Cell Physiol 131:402–408

Waller BF, Gorfinkel HJ, Rogers FJ, Kent K, Roberts W (1984) Early and late morphological changes in major epicardial coronary arteries after percutaneous transluminal coronary angioplasty. Am J Cardiol 53 (Suppl C):42C–47C

Weich HA, Iberg N, Klagsbrun M, Folkman J (1990) Expression of acidic and basic fibroblast growth factors in human and bovine vascular smooth muscle cells. Growth Factors 2:313–320

Wight LN (1989) Cell biology of arterial proteoglycans. Arteriosclerosis 9:1–20

Wilcox JN, Smith KM, Williams LT, Schwartz SM, Gordon D (1988) Platelet-derived growth factor mRNA detection in human atherosclerotic plaque by in situ hybridization. J Clin Invest 82:1134–1143

Willerson JT, Golino P, Eidt J, Campbell WB, Buja LM (1989) Specific platelet mediators and unstable coronary artery lesions. Circulation 80:198–205

Winkles JA, Friesel R, Burgess WH, Howk R, Mehlman T, Weinstein R, Maciag T (1987) Human vascular smooth muscle cells both express and respond to heparin-binding growth factor 1 (endothelial cell growth factor). Proc Natl Acad Sci, USA 84:7124–7128

Chapter 5

The Monocyte and Endothelial Injury in Atherogenesis

R. G. Gerrity

Monocytes and Lesion-Susceptible Areas

The macrophage foam cell has been recognized as a hallmark of the atherosclerotic plaque for over a century. However, the potential importance of the blood monocyte in atherosclerosis was first pointed out by Leary as late as 1941. Poole and Florey (1958) later discussed the blood as a source of lesion foam cells in hypercholesterolemic rabbits, and demonstrated a macrophage traversing the endothelium. Still and O'Neal (1962) were the first to demonstrate, ultrastructurally, the adherence to and penetration of the endothelium by monocytes in hypercholesterolemic animals, a finding subsequently described by numerous others under a wide variety of conditions (Jerome and Lewis 1984, 1985; Joris et al. 1983; Gerrity 1981a, 1981b). However, until the last decade, only Kim et al. (1966) attempted to establish the circulating monocyte/macrophage as a major factor in atherogenesis, and it was unclear whether lipid accumulation in lesion macrophage foam cells occurred due to sequestration of blood lipophages in the arterial wall, or by the accumulation of intimal lipid by monocytes which had previously traversed the endothelium. Furthermore, the "response to injury" hypothesis in its initial format (Ross and Glomset 1976) did not define any potential role of the blood monocyte in lesion initiation, and the relationships, if any, between endothelial injury and monocyte penetration of the intima remained unclear.

Extensive studies in the past decade on the role of the monocyte in atherosclerosis have shown, firstly, that monocyte involvement in lesion development is one of the earliest cellular responses (Jerome and Lewis 1984, 1985; Joris et al. 1983; Gerrity, 1981a, 1981b) and, secondly, that the response of the monocyte is targeted to specific sites in the arterial wall, as opposed to a generalized involvement (Gerrity 1981a, 1981b). Thus, any consideration of the relationships between the circulating monocyte and the arterial endothelium must take into account that atherosclerotic lesions do not occur randomly in the arterial system. Rather, they have been shown to favor certain regions, while sparing others, both in the human (McGill 1968a, 1968b; Mitchell and Schwartz 1965) and in experimental models (Jerome and Lewis 1984; Gerrity 1981a, 1981b; Gerrity et al. 1979). Computer-assisted image-processing techniques have recently been used to quantify statistically the topographical distribution of lesions (Cornhill et al., 1985), and have defined areas which are susceptible (S) and non-susceptible (NS) to atherosclerotic lesion formation. In animal models, at least, monocyte recruitment occurs preferentially in S-regions, and results in the formation of fatty streak lesions (Gerrity 1981a, 1981b). In the swine model, such lesions in the thoracic aorta seldom progress beyond the fatty streak, whereas those in the abdominal aorta progress to form advanced fibrous plaques (Gerrity 1981b). In this model, then, fatty streaks may be precursors to fibrous plaques in some arterial S-regions, but not in others. Similarly, Stary (1985) has demonstrated that lesions in very young humans consist predominantly of macrophage foam cells, and that advanced fibrous plaques can be found at the same anatomical sites in older individuals. If, as these data suggest, the fatty streak is a precursor to the clinically significant fibrous plaque, then the mechanisms controlling monocyte recruitment and monocyte–endothelial cell interactions in S-areas are of great importance, since the monocyte-derived macrophage foam cell is the predominant cell type in the fatty streak lesion.

Morphology of Lesion-Susceptible Areas

The interactions between circulating monocytes and the endothelium overlying S-regions leading to specific monocyte recruitment must be considered in the broader concept of the altered structure and function of the arterial wall in S-regions. Adherence and penetration of S-area endothelium by monocytes do not appear to be associated with overt endothelial damage or denudation, at least in animal models (Jerome and Lewis 1984, 1985; Joris et al. 1983; Gerrity 1981a, 1981b). However, the endothelium and intima of lesion-susceptible areas differ structurally from those of overlying adjacent areas which are not susceptible to early lesion formation, even in normal animals. In the latter, endothelial cells are consistently elongated, flattened, and oriented in the direction of blood flow. In contrast, S-area endothelium exhibits an almost cuboidal structure in both surface and sectional view (Gerrity et al. 1979). Although early light microscopic studies (Caplan et al. 1974) were unable to detect, quantitatively, differences in cell shape and size between the two areas, subsequent ultrastructural quantitative studies (Gerrity et al. 1977; Gerrity and Naito 1980) using perfusion-fixed vessels showed that endothelial cells in lesion-susceptible areas are larger and more rounded than those in NS-areas (surface areas $741 \pm 40\,\mu m^2$ vs. $545 \pm 38\,\mu m^2$; length:width ratio 1.39 ± 0.03 vs.

3.30 ± 0.09, respectively). Moreover, these parameters in NS-areas can be made to duplicate those in S-areas by altering hemodynamic and flow conditions through partial coarctation of the vessel distal to the area (Gerrity and Naito 1980). These differences in cell size in lesion-susceptible areas are of particlular interest with respect to human vessels in that Repin et al. (1984) have reported that the arterial endothelium of infants exhibits smaller cells and greater cell density than that of uninvolved adult endothelium. Moreover, they demonstrated that the size of endothelial cells over fatty streaks and plaques, like that of swine aortic susceptible areas, is greater than that of uninvolved areas or infant vessels, with a corresponding decrease in cell density. Giant and large cells covered up to 41% of plaque surfaces, and small cells occupied only 24% of the entire surface. In contrast, in normal infant vessels, 75% of the surface is covered by small cells. Thus, although S-areas do not show morphologically demonstrable damage, well-defined alteration of endothelial structure has been described in these regions in animal models, which may be analogous to similar alterations seen in endothelium overlying plaques in humans.

Endothelial Transport and Intimal Accumulation in S-Areas

Whereas receptor-mediated endocytosis is considered to be the chief mechanism by which specific macromolecules are taken up for metabolism by the cell, it is primarily non-receptor-mediated bulk-phase pinocytosis by the endothelium which results in the passage of blood macromolecules through the intact endothelial layer into the vessel wall (Simionescu et al. 1973, 1975). This process is thus of considerable interest with respect to the transport and accumulation of plaque lipids, particularly cholesterol, from the lipoproteins of the blood into lesion-susceptible sites. Studies using intravenously injected labelled molecules (Bell et al. 1974a, 1974b) or electron-dense probes such as ferritin (Gerrity and Schwartz 1977) have provided considerable insight into this problem, and have demonstrated that lesion-susceptible areas exhibit enhanced intimal accumulation of blood macromolecules.

Such studies in young, normal swine (Gerrity and Schwartz 1977) have shown that the numbers of endothelial pinocytotic vesicles are similar in lesion-susceptible and non-susceptible areas, and that the carrying capacity of vesicles for electron-dense probes such as ferritin is likewise the same. In older swine, NS-area endothelial cells exhibit more vesicles per unit sectional area than adjacent S-areas. Despite this age-related change, which would favor greater intimal accumulation in NS-areas, approximately two to three times as many vesicles take up intravenously injected ferritin in S- compared to NS-areas (Gerrity and Schwartz 1977). As a result, more ferritin accumulates in the intima of such areas in normal animals, and such accumulation is further enhanced under conditions of hyperlipidemia, particularly by intracellular uptake of the probe by macrophages. Thus, the expanded intima of lesion-susceptible areas may provide a "sink" for enhanced accumulation of molecules taken up by bulk-phase vesicular transport. It is of considerable importance to note that hyperlipidemia does not increase the percentage of vesicles carrying ferritin, which is two to three times greater in lesion-susceptible areas compared to non-susceptible areas. However, intimal accumulation is increased by hyperlipemia both extracellularly, and to an even greater extent, within macrophages (Gerrity 1990). Lesion-susceptible areas demonstrate increased accumulation of albumin (Bell et al. 1974a), fibrinogen (Bell et al. 1974b), cholesterol (Somer and Schwartz 1972) and lipoprotein (Hoff et al. 1983; Feldman et al. 1984) in addition to ferritin, supporting the hypothesis that this property is due to non-specific bulk-phase vesicular transport across the endothelium.

Of interest in these studies is that ferritin was never observed in the shorter, sometimes vacuolated junctional regions of lesion-susceptible endothelium, and secondly, in those rare cases in which focal endothelial denudation was observed, almost no ferritin accumulation was observed in the exposed intima. These findings are consistent with those of Minick et al. (1979), who showed that intimal lipoprotein accumulation occurred subsequent to re-endotheliatization in ballooned arteries, but did not occur in exposed intima. More recent studies in swine have demonstrated, both structurally (Feldman et al. 1984) and biochemically (Hoff et al. 1983), that lesion-susceptible areas preferentially accumulate apo B-containing lipoproteins even in normal animals, a condition exaggerated by hyperlipemia. With prolonged hyperlipemia, however, even non-susceptible areas accumulate lipoprotein (Hoff et al. 1974). Since these lipoproteins are of a molecular size necessitating vesicular transport, it is highly likely that the preferential lipoprotein accumulation in susceptible areas is a result of the greater bulk-phase vesicular transport and enhanced accumulation in the thickened intima demonstrated by the ferritin studies. Hoff, in a series of papers, has demonstrated the presence of apo B-containing lipoproteins in a wide variety of human arteries (Hoff et al. 1974, 1975a, 1975b, 1975c, 1980). Although this accumulation has not been linked to specific predilection sites, it is significant that he has demonstrated the presence of apo B-containing lipoproteins even in grossly normal vessels (Hoff et al. 1975c). If, as suggested by studies of ferritin and lipoprotein accumulation in swine, lesion-susceptible areas preferentially accumulate lipoprotein through inherent differences in bulk-phase transport, coupled with enhanced intimal "trapping", then the concept of endothelial injury as an initiating factor in atherogenesis must be re-examined. Enhanced bulk-phase transport, as well as increased accumulation of a wide variety of large blood macromolecules (Bell et al. 1974a, 1974b; Gerrity and Schwartz 1977; Gerrity 1990; Somer and Schwartz, 1972; Feldman et al. 1984) occurs in these areas even in normal swine. Thus the preferential accumulation of atherogenic lipoproteins in such regions is not the result of endothelial injury induced by hyperlipemia, but rather, the superimposition of altered blood composition on areas of normally altered structural and functional characteristics which are independent of induced injury. The resultant focal accumulation of atherogenic macromolecules may then trigger the events leading to lesion formation. The focal nature of lesion formation in human atherosclerosis (Cornhill 1986) would suggest the existence in human arteries of areas analogous to S-areas in swine. The possibility therefore exists that human lesions may also initiate through a similar mechanism. Certainly, at the very least, the results of animal studies would suggest that we must look for more subtle forms of endothelial "injury" in the form of altered function as an initiating factor in atherogenesis.

Monocyte Recruitment in Lesion-Susceptible Areas

It is clear from these studies that monocytes are preferentially recruited into S-areas of hyperlipemic animal models (Jerome and Lewis 1984, 1985; Joris et al. 1983; Gerrity 1981a, 1981b) and that such areas demonstrate altered endothelial cell and intimal structure, permeability and intimal accumulation of blood macromolecules, including lipoproteins and altered metabolic processes. However, the relationships between these features and the mechanisms which initiate and modulate monocyte adhesion and intimal penetration are poorly understood. Chronologically, in animal models, monocyte involvement can be documented after very short periods of hyperlipemia (Gerrity 1981a) at stages concurrent with intimal accumulation of apo B-containing lipoproteins (Hoff et al. 1983; Feldman et al. 1984) and prior to the appearance of grossly visible lesions. There is evidence that such involvement occurs preferentially at the periphery of developing lesions (Jerome and Lewis 1984), and that adherence and intimal migration of monocytes becomes more focal as lesions develop (Gerrity 1981b), suggesting highly localized control mechanisms. The preferential and focal nature of monocyte recruitment in S-areas would suggest a chemotactic response as a likely controlling mechanism, since chemotactic migration is directional, gradient dependent and receptor mediated (Snyderman and Friedman 1980). Monocytes are known to respond chromatically to numerous stimulatory molecules potentially associated with the arterial wall, including thrombin (Bar-Shavit et al. 1983), platelet-derived growth factor (PDGF) (Deuel et al. 1982), fibronectin (Norris et al. 1982), kallikrein and plasminogen activator (Gallin and Kaplan 1974), elastin and collagen fragments (Hunninghake et al. 1981), various inflammatory factors (Snyderman et al. 1971), and monocyte colony-stimulating factor (Rajavashisth et al. 1990). A chemotactic factor specific for monocytes from hyperlipemic animals has been demonstrated in S-regions of hyperlipemic swine (Gerrity et al. 1985), and a 14.4-kDa monocyte chemotactic protein (MCP-1) has been shown to be produced by cultured aortic smooth muscle cells (Jauchem et al. 1982) and endothelial cells (Berliner et al. 1986), as well as by several tumor cell lines (Graves et al. 1989). MCP-1 has been shown to account for all the chemotactic activity produced by smooth muscle and endothelial cells in culture, and production of MCP-1 by these cells can be induced by minimally oxidized LDL (MM-LDL) (Cushing et al. 1990). MM-LDL, which is recognized by the LDL-, but not the scavenger-receptor, also induces endothelial cells to produce (Rajavashisth et al. 1990) monocyte colony-stimulating factor (M-CSF), which not only enhances production of monocytes by bone marrow progenitor cells but also elicits a chemotactic response from monocytes. It is of considerable interest that aortic endothelial cells produce M-CSF, in that hyperlipemic swine demonstrate a marked monocytosis at times concomitant with early atherogenesis and monocyte recruitment (Averill et al. 1989). This monocytosis, which can result in as much as a doubling in the number of circulating monocytes, appears to result from increased proliferation of monocytic progenitor cells in the bone marrow. Bone marrow of hyperlipemic swine contains significantly elevated numbers of bone marrow progenitor cells (BMC) compared to controls, which, when placed in culture, preferentially differentiate into monocyte colonies, even in the absence of exogenous colony-stimulating factors (Averill et al. 1989).

Taken in combination, these two phenomena result in a significantly greater total number of monocytic progenitor cells in hyperlipemic swine compared to age- and sex-matched controls. This enhanced monocyte proliferation in culture would suggest that prior activation of monocyte progenitor cell proliferation by factors present in the hyperlipemic environment has occurred in vivo. Furthermore, serum from hyperlipemic swine preferentially stimulates proliferation of bone marrow monocyte progenitor cells in vitro, regardless of whether the bone marrow source is from hyperlipemic or normal swine. This finding suggests that increased levels of monocyte (M) or granulocyte/monocyte (GM) CSF may be present in HL-swine sera. M-CSF in other systems is a glycoprotein required for the proliferation of monocyte precursor cells (Metcalf 1977). In the present case, M-CSF may be produced or secreted by cells in vivo in response to a factor in HL-swine sera. Alternatively, factors such as modified lipoproteins may activate CSF-producing lesion cells in vivo to produce or secrete enhanced levels of M-CSF, block the production of factors inhibitory to bone marrow progenitor cell proliferation, or act directly on progenitor cells preferentially to produce monocytes. Regardless of the mechanism of action, the results of these experiments demonstrate that HL-swine sera elicit enhanced monocytic colony formation in N- as well as in HL-swine BMC cultures. Thus, prior activation or alteration of BMC in a hyperlipemic environment is not necessary for enhanced monocytic progenitor cell proliferation in vivo in response to factors in HL-swine sera. The results also demonstrate that BMC from HL-swine possess an enhanced intrinsic capability to proliferate and form monocytic colonies compared to N-swine BMC, regardless of whether the stimulus is HL- or N-swine sera. Taken together, these results indicate that increased BMC proliferation resulting in the observed monocytosis may not require long-term pre-activation of monocytic progenitor cells by factors in the hyperlipemic environment. However, once activated, BMC retain their increased capacity to generate elevated numbers of monocytic progenitor cells. In other words, residence of bone marrow stem cells within the hyperlipemic environment has led to a significant increase in the proportion and/or sensitivity of monocytic progenitor cells to proliferate in response to circulating in vivo factor(s). This change is not altered following removal of BMC from the hyperlipemic environment. These findings also have implications for plaque regression. If increased capacity for monocyte production is retained after removal of hyperlipemic conditions, a persisting monocytosis could significantly contribute to removal of lipid from the plaque, as previously postulated (Gerrity 1981b). In fact, the presence of large numbers of monocyte/macrophages in regressing lesions has previously been documented in the swine model (Daoud et al. 1981, 1985).

An essential primary event for monocyte migration into S-regions or lesions is adhesion of the circulating monocyte to the endothelial surface. The mechanism controlling this area-specific adhesion is as yet unclear. As discussed above, although overt endothelial damage does not appear to be a major factor in these initiating stages of atherogenesis, altered endothelial cell structure and function are present in such areas, resulting, at least, in accumulation of lipoproteins, which, when modified, have been shown to elicit the production of chemotactic proteins and growth factors by endothelial cells in culture. It is equally feasible that such molecules may elicit production (either directly or indirectly mediated) of leukocyte adhesion molecules similar to those believed to be active in various inflammatory and immune responses (see Cotran and Pober 1990, for review). Several such adhesion molecules have been identified, and appear to be regulated by various cytokines (Cotran and Pober 1990; Pober and Cotran 1990). Whether these or other adhesion molecules are active in monocyte adhesion to endothelium in atherogenesis is unknown, but it is of interest that interleukin-1 (IL-1) induces synthesis of MCP-1 in cultured endothelium, and also promotes three- to sixfold and twentyfold increases in monocyte and granulocyte adherence, respectively (Reidy 1990).

In summation, evidence to date indicates that monocyte recruitment into specific lesion-susceptible sites in arteries is one of the earliest cellular events detectable in the atherogenic sequence. This process appears to occur in the absence of overt endothelial injury, and is probably mediated by the production of specific attraction molecules by lesion cells. The endothelium overlying susceptible areas exhibits structural and functional characteristics unique to such sites even in normal animals. Studies in tissue culture indicate the possibility that lipoproteins or other blood macromolecules (which are known to accumulate in lesions) may provide the stimulus which induces production of adhesion and migratory molecules regulating monocyte recruitment. Complex feedback mechanisms may exist between lesion cells and monocyte progenitor cells to augment the number of circulating monocytes. It is evident that endothelial cells are capable of producing a wide variety of adhesive, chemotactic, thrombogenic and mitogenic compounds in culture. A major challenge will be to determine which of these factors play a role in the atherogenic sequence. At the very least, it is evident that our perception of endothelial "injury" must be redefined in terms of functional alterations either pre-existing in S-regions or induced by hyperlipemia.

References

Averill LE, Meagher RC, Gerrity RG (1989) Enhanced monocyte progenitor cell proliferation in bone marrow of hyperlipemic swine. Am J Pathol 135:369

Bar-Shavit R, Kahn A, Fenton JW, Wilner GD (1983) Chemotactic responses of monocytes to thrombin. J Cell Biol 96:282

Bell FP, Adamson IL, Schwartz CJ (1974a) Aortic endothelial permeability to albumin: focal and regional patterns of uptake and transmural distribution of ^{131}I-albumin in the young pig. Exp Mol Pathol 20:57

Bell FP, Gallus AS, Schwartz CJ (1974b) Focal and regional patterns of uptake and the transmural distribution of ^{131}I-fibrinogen in the pig aorta in vivo. Exp Mol Pathol 20:281

Berliner JA, Territo M, Almada L, Carter A, Shafonsky E, Fogelman AM (1986) Monocyte chemotactic factor produced by large vessel endothelial cells in vitro. Arteriosclerosis 6:254

Caplan BA, Gerrity RG, Schwartz CJ (1974) Endothelial cell morphology in focal areas of in vivo Evans Blue uptake in the young pig aorta. I. Quantitative light microscope findings. Exp Mol Pathol 21:102

Cornhill JF (1986) Topographic probability mapping of atherosclerosis. Report 763813/715703, Ohio State University Research Foundation (NHLBI, N01-HV-38019)

Cornhill JF, Barrett WA, Herderick EE, Mahley RW, Fry DL (1985) Topographic study of sudanophilic lesions in cholesterol-fed minipigs by image analysis. Arteriosclerosis 5:415

Cotran RS, Pober JS (1990) Cytokine-endothelial interactions in inflammation, immunity, and vascular injury. J Am Soc Nephrol 1:225

Cushing SD, Berliner JA, Valente AJ, Territo MC, Navab M, Parhami F, Gerrity RG, Schwartz CJ, Fogelman AM (1990) Minimally modified low density lipoprotein induces monocyte chemotactic protein 1 in human endothelial cells and smooth muscle cells. Proc Natl Acad Sci USA 87:5134

Daoud AS, Jarmolych J, Augustyn JM, Fritz KE (1981) Sequential morphologic studies of regression of advanced atherosclerosis. Arch Pathol Lab Med 105:233

Daoud AS, Fritz KE, Jarmolych J, Frank AS (1985) Role of macrophages in regression of atherosclerosis. Ann NY Acad Sci 454:101

Deuel TF, Senior RM, Haung JS, Griffin GL (1982) Chemotaxis of monocytes and neutrophils to platelet-derived growth factor. J Clin Invest 69:1046

Faggiotto A, Ross R, Harker L (1984) Studies of hypercholesterolemia in the nonhuman primate. I. Changes that lead to fatty streak formation. Arteriosclerosis 4:323

Feldman DL, Hoff HF, Gerrity RG (1984) Immunohistochemical localization of apoprotein B in aortas from hyperlipemic swine: preferential accumulation in lesion-prone areas. Arch Pathol Lab Med 108:817

Gallin JI, Kaplan AP (1974) Mononuclear cell chemotactic activity of kallikrein and plasminogen activator and inhibition by CI inhibitor and α-macroglobulin. J Immunol 129:1612

Gerrity RG (1981a) The role of the monocyte in atherogenesis. I. Transition of blood-borne monocytes into foam cells in fatty lesions. Am J Pathol 102:181

Gerrity RG (1981b) The role of the monocyte in atherogenesis. II. Migration of foam cells from atherosclerotic lesions. Am J Pathol 103:191

Gerrity RG (1990) Arterial endothelial structure and permeability as it relates to susceptibility to atherogenesis. In: Glagov S, Newman III WP, Schaffer SA (eds) Pathobiology of the human atherosclerotic plaque. Springer-Verlag, New York, pp 13–45

Gerrity RG, Naito KH (1980) Alteration of endothelial cell surface morphology after experimental aortic coarctation. Artery 8:267

Gerrity RG, Schwartz CJ (1977) Structural correlates of arterial endothelial permability in the Evans Blue model. In: Sinzinger H, Auerswald WA, Jellinek H, Feigl W (eds) Prog Biochem Pharmacol, S. Karger, Basel, 13:134–137

Gerrity RG, Richardson M, Bell FP, Somer JB, Schwartz CJ (1977) Endothelial cell morphology in areas of in vivo Evans Blue uptake in the young pig aorta. II. Ultrastructure of the intima in areas of differing permeability to proteins. Am J Pathol 89:313

Gerrity RG, Naito HK, Richardson M, Schwartz CJ (1979) Dietary-induced atherogenesis in swine. I. Morphology of the intima in pre-lesion stages. Am J Pathol 95:775

Gerrity RG, Goss JA, Soby L (1985) Control of monocyte recruitment by chemotactic factor(s) in lesion-prone areas of swine aorta. Arteriosclerosis 5:55

Graves DT, Jiang YL, Williamson MJ, Valente AJ (1989) Identification of monocyte chemotactic activity produced by malignant cells. Science 245:1490

Hoff HF, Jackson RI, Mao JT, Gotto AM (1974) Localization of low density lipoproteins in arterial lesions from normolipemics employing a purified fluorescent labeled antibody. Biochim Biophys Acta 351:407

Hoff HF, Heideman CI, Noon JP, Meyer JS (1975a) Localization of apolipoproteins in human carotid artery plaques. Stroke 6:531

Hoff HF, Heideman CI, Gaubatz JW (1975b) Apo-low density lipoprotein localization in intracranial and extracranial atherosclerotic lesions from human normolipoproteinemics and hyperlipoproteinemics. Arch Neurol 32:600

Hoff HF, Lie JT, Titus JL, Jackson RL, DeBakey ME, Bayardo R, Gotto AM (1975c) Lipoproteins in atherosclerotic lesions: localization by immunofluorescence of apo-low density lipoproteins in human atherosclerotic arteries from normal and hyperlipoproteinemics. Arch Pathol 99:253

Hoff HF, Ruggles BM, Bond MG (1980) A technique for localizing LDL by immunofluorescence in formalin-fixed and paraffin-embedded atherosclerotic lesions. Artery 6:328

Hoff HF, Gerrity RG, Naito HK, Dusek D (1983) Quantitation of apoliopoprotein B in aortas of hypercholesterolemic swine. Lab Invest 48:492

Hunninghake GW, Davidson JM, Rennard S, Szapiel S, Gadek JR, Crystal RG (1981) Elastin fragments attract macrophage precursors to diseased sites in pulmonary emphysema. Science 212:925

Jauchem JR, Lopez M, Sprague EA, Schwartz CJ (1982) Mononuclear cell chemoattractant activity from cultured arterial smooth muscle cells. Exp Mol Pathol 37:166

Jerome WG, Lewis JC (1984) Early atherogenesis in white carneau pigeons. I. Leukocyte margination and endothelial alterations at the celiac bifurcation. Am J Pathol 116:56

Jerome WG, Lewis JC (1985) Early atherogenesis in white carneau pigeons. II. Ultrastructural and cytochemical observations. Am J Pathol 119:210

Joris I, Nunnari T, Krolikowski JJ, Majno FJ (1983) Studies on the pathogenesis of atherosclerosis. I. Adhesion and emigration of mononuclear cells in the aorta of hypercholesterolemic rats. Am J Pathol 113:341

Kim HS, Suzuki M, O'Neal RM (1966) The lipophage in hyperlipemic rats: an electron microscopic study. Exp Mol Pathol 5:1

Leary T (1941) The genesis of atherosclerosis. Arch Pathol 32:507

McGill HC Jr (1968a) Persistent problems in the pathogenesis of atherosclerosis. Atherosclerosis 4:443

McGill HC Jr (1968b) The geographic pathology of atherosclerosis. Williams and Wilkins, Baltimore

Metcalf D (1977) Neutrophil and macrophage colony formation by normal cells. Recent Results Cancer Res 61:56

Minick CG, Stemmerman MB, Insul W (1979) Role of endothelium and hypercholesterolemia in intimal thickening and lipid accumulation. Am J Pathol 95:131

Mitchell JRA, Schwartz CJ (1965) Arterial disease. Blackwell Scientific Publications, Oxford

Norris DA, Clark RAF, Swigart LM, Huff JC, Weston WL, Howell SE (1982) Fibronectin fragment(s) are chemotactic for human peripheral blood monocytes. J Immunol 129:1612

Pober JS, Cotran RS (1990) Cytokines and endothelial cell biology. Physiol Rev 70:427

Poole JCF, Florey HW (1958) Changes in the endothelium of the aorta and behaviour of macrophages in experimental atheroma of rabbits. J Pathol Bacteriol 75:245

Rajavashisth TB, Andalibi A, Territo MC, Berliner JA, Navab M, Fogelman AM, Lusis AJ (1990) Nature (Lond) 344:254

Reidy MA (1990) In vivo endothelial injury. In: Subbiah MTR (ed) Atherosclerosis: a pediatric perspective. CRC Press, Boca Raton, Florida, pp 31–42

Repin VS, Dolgov VV, Zaikina OE, Novikov ID, Antonov AS, Nikolaeva MS, Smirnov VN (1984) Heterogeneity of endothelium in human aorta: a quantitative analysis by scanning electron microscopy. Atherosclerosis 50:35

Ross R, Glomset JA (1976) The pathogenesis of atherosclerosis. N Engl J Med 295:369

Simionescu N, Simionescu M, Palade GE (1973) Permeability of muscle capillaries to exogenous myoglobin. J Cell Biol 57:424

Simionescu N, Simionescu M, Palade GE (1975) Permeability of muscle capillaries to small heme-peptides: evidence for the existence of patent transendothelial channels. J Cell Biol 64:586

Snyderman R, Friedman EJ (1980) Demonstration of a chemotactic factor receptor on macrophages. J Immunol 124:2754

Snyderman R, Shin HS, Hausman MS (1971) A chemotactic factor for mononuclear leukocytes. Proc Soc Exp Biol Med 138:287

Somer JB, Schwartz CJ (1972) Focal [^{3}H]-cholesterol uptake in the pig aorta. II. Distribution of [^{3}H]-cholesterol across the aortic wall in areas of high and low uptake in vivo. Atherosclerosis 16:377

Stary HC (1985) Evolution and progression of atherosclerosis in the coronary arteries of children and adults. In: Bates SR, Gangloff EC (eds) Atherosclerosis and aging. Springer-Verlag, Heidelberg, p 20

Still WJS, O'Neal RM (1962) Electron microscopic study of experimental atherosclerosis in the rat. Am J Pathol 40:21

Chapter 6

Lp [a]: A Lipoprotein Class with Atherothrombotic Potential

A. M. Scanu

Introduction

Lipoprotein [a] or Lp [a] represents a class of lipoprotein particles associated with an increased prevalence of atherosclerotic cardiovascular disease (ASCVD) (Utermann 1989; Scanu and Fless 1990; Berg 1990). At this time this notion rests predominantly on epidemiological data although emerging experimental evidence is now permitting to formulate hypotheses on the possible mechanism(s) of this pathogenicity. This experimental evidence will be reviewed and correlated with the current understanding of the structural properties of Lp [a].

Structural Properties of Lp [a] (McLean et al. 1987; Fless et al. 1990; Scanu and Fless 1990)

The unique structural feature of Lp [a] is the apo B-100-apo(a) complex, the protein moiety of this lipoprotein particle (Table 6.1). Apo B-100 of Lp [a] appears to be structurally very similar if not identical to apo B-100 of authentic low-density lipoproteins (LDL). Apo B-100 is linked by a disulfide bridge to apo(a), the specific glycoprotein of Lp [a]. The apo B-100–apo(a) complex, freed of lipids, is hydrophilic (contrary to apo B-100) and also lipophilic. This latter capacity permits apo B-100–apo(a) to associate with both LDL-like cholesterol-rich particles, or CE-Lp [a], and triglyceride-rich VLDL-like

Table 6.1. Properties of the apo B-100–apo(a) complex, the protein moiety of Lp[a]

Apo B-100–apo(a) molar ratio usually 1:1, occasionally 1:2
Apo B-100 is linked to apo(a) by a disulfide bridge
Chemical linkage occurs in the liver
Apo B-100–apo(a) is commonly affiliated with LDL-like cholesteryl ester-rich particles or CE-Lp[a]
Apo B-100–apo(a) can also affiliate with VLDL-like triglyceride rich particles, or TG-Lp[a]
Apo B-100–apo(a) in its lipid-free form is water soluble

---

particles, or TG-Lp[a]. One of the characteristics of Lp[a] is to be heterogeneous in density and in size. This heterogeneity is mostly dependent on the variable amount of lipid that the apo B-100–apo(a) complex carries, but also on the apo(a) size polymorphism that varies from 300 to 800 kDa. The apo B-100–apo(a) molar stoichiometry is usually 1:1 but in some cases is 1:2. A correlation exists between size of the apo B-100–apo(a) complex and Lp[a] hydrated density. For instance, the apo B-100–apo(a) complex of molecular weight about 800 kDa – 500 kDa from apo B-100 and 300 kDa from apo(a) – preferentially affiliates with the Lp[a] particles of the highest buoyancy whereas the apo B-100–apo(a) complex of molecular weight about 1300 kDa – 500 kDa from apo B-100 and 800 kDa from apo(a) – affiliates preferentially with the heavier species of Lp[a]. The mechanism underlying this preferential affiliation is unknown. The carbohydrate component of apo(a), about 30% by weight, might also contribute to the size heterogeneity of apo(a) but to a much lesser extent than that contributed by the polypeptide chain. The recent application of molecular biology techniques has permitted one to establish that the size heterogeneity of apo(a) is due to differences in the number of kringle 4 domains, named because of their high degree of homology with the kringle 4 of plasminogen. Apo(a) lacks kringles 1, 2 and 3 that are present in plasminogen; it contains one copy of kringle 5 and the same protease region as plasminogen. However, apo(a) is incapable of clot lysis due to the fact that it lacks the activation site that plasminogen has. Since apo(a) has several kringle 4 repeats contrary to the only one in plasminogen, at least on theoretical grounds, apo(a) should be able to compete with the function of plasminogen in the fibrinolytic system.

Lp[a] Physiology (Utermann 1989; Scanu 1990; Scanu and Fless 1990)

Up to 80% of the Caucasian population has levels of plasma Lp[a] that are in the limits of normal, i.e. below 5–7 mg/dl (in terms of protein) or below 25–30 mg/dl (in terms of whole Lp[a]). The remainder of the population has higher plasma Lp[a] that have been associated with an increased prevalence of ASCVD. Essentially all subjects have Lp(a) in their plasma, with an inter-individual variation up to 1000 times. The factors that control the plasma levels of Lp[a] are not clearly established. The liver is the major if not the only organ of Lp[a] synthesis that is under the control of the apo(a) gene and its several alleles. The factors controlling the expression of this gene and the extent to which it contributes to the maintenance of the plasma levels of Lp [a] have not been clearly established (Utermann 1989). Catabolism is also likely to have a role in this regard. Unfortunately, the information on the subject is conflicting (Utermann 1989; Scanu and Fless 1990); however, the participation of the LDL receptor pathway, although suggested by some workers, is not probable. The same can be said about the scavenger receptor pathway, except for chemically modified Lp [a], for instance by malondialdehyde (Haberland et al. 1989). Evidence for a role by anon-receptor-mediated pathway has recently been obtained (Snyder et al. 1990) and there is also experimental evidence favoring the transport of intact Lp [a] from plasma to the extravascular compartment via the vascular endothelium (Rath et al. 1989; Cushing et al. 1989). With so many uncertainties it is probably premature to ask whether "normal" levels of plasma Lp [a] have a physiological meaning. The lack of factual information on the subject invites only speculation: Lp [a] might be involved in a selective cholesterol delivery role for tissue repair and/or in the modulation of the activity of the components of the fibrinolytic system at the endothelial surface. A useful role for low plasma levels of Lp [a] would speak in favor of an evolutionary advantage for Lp [a]. This would be in keeping with the presence of Lp [a] in the plasma of high animal species like apes and Old World monkeys, with the outstanding exception of the hibernating hedgehog, for which the philogeny of Lp [a] remains a mystery (Utermann 1989; Scanu and Fless 1990; Scanu 1990).

Lp [a] as a Cardiovascular Pathogen

Keeping in mind that high plasma levels of Lp [a] have been associated with an increased prevalence of ASCVD (Berg 1990) the mechanisms dealing with the atherogenic and thrombogenic potentials of this lipoprotein will be considered separately, although they may be operative in a synergistic way.

Atherogenic Potential (Scanu and Fless 1990; Scanu 1990)

This might be due to the LDL-like properties of Lp [a] by a process that would first involve the transfer of Lp [a] from the circulation to the subendothelial intima of the arterial wall. Here Lp [a] would undergo chemical modifications and subsequently be taken up by the scavenger receptor of resident macrophages that are then transformed into foam cells as a first step towards the formation of the atherosclerotic plaque (Table 6.2). This hypothesis views the

Table 6.2. Postulated atherogenicity of Lp [a]

Lp [a] traverses the vascular endothelium and reaches the intima as an intact particle
In the intima Lp [a] is structurally modified by the action of oxygen-free radicals, matrix components (proteoglycans, glycosaminoglycans, etc.) or fibrin
Modified Lp [a] is taken up by resident macrophages that then become foam cells due to entrapment of Lp [a] cholesterol

---

endothelium as a filter, the driving force for the transport being the Lp[a] gradient between the plasma and the arterial wall. This hypothesis also assumes that in the intima Lp[a] would undergo chemical modification. Although plausible, it is likely that this hypothesis is rather simplistic in that the endothelium may not act as a simple filter but respond to the high concentrations of Lp[a] and undergo structural and functional changes which by themselves might influence Lp[a] permeability. This permeability may also be influenced by focal sites of endothelial inflammation or injury. This would provide an explanation for our recent findings that subjects with an extensive coronary artery involvement (three-vessel disease) have normal or subnormal levels of plasma Lp[a] (Sorrentino et al. 1990).

Thombogenic Potential (Loscalzo 1990; Scanu and Fless 1990)

In vitro and ex vivo studies have clearly shown that apo(a) can compete for plasminogen in key steps of the fibrinolytic process, for instance binding to fibrin(ogen) and fragments as well as in tissue plasminogen-dependent activation of plasminogen to plasmin. In keeping with these findings, Lp[a] has been shown to attenuate clot lysis (Loscalzo 1990). Lp[a] can also compete with plasminogen for its binding to the plasminogen receptor, resulting in the availability of receptor-immobilized plasminogen better suited for activation to plasmin (Table 6.3). These competitive activities by Lp[a] should occur at the

Table 6.3.   Postulated thrombogenicity of Lp[a]

Lp[a] competes for the binding of plasminogen with fibrin(ogen) or fragments
Lp[a] interferes with the process of plasminogen → plasmin conversion
Lp[a] attenuates clot lysis
Lp[a] competes for the binding of plasminogen to the plasminogen receptor

level of the kringle 4 domains although the actual kringle(s) involved and the precise site(s) within each kringle remain to be established. It should be noted, however, that in spite of the experimental evidence thus far obtained, a convincing documentation concerning an Lp[a]-dependent thrombogenic state in living subjects is not yet available.

Considerations on the Thrombo-atherogenicity of Lp[a]

The postulated thrombo-atherogenic potential of Lp[a] invites one to ask whether an elevation of plasma Lp[a] alone is sufficient to predispose or even cause ASCVD. At this time, this may only apply to subjects with exceptionally high plasma levels of Lp[a], i.e. of the order 100–200 mg/dl. However, the incidence of this high degree of hyperLp[a] proteinemia is infrequent since, most commonly, one is confronted with patients having levels of Lp[a] in the range of 8–25 mg/dl (in terms of protein) or 30–80 mg/dl (in terms of total lipoprotein) where the pathogenic action of Lp [a] is accentuated by the presence of other dyslipoproteinemias, i.e. high LDL and/or low HDL in addition to a family history of ASCVD. In those cases it is difficult to identify the respective role of each of these factors, although it is reasonable to assume that they may have additive or even multiplicative effects (Utermann 1989). For instance, it has been recently shown that patients with heterozygous familial hypercholester-olemia and hyperLp [a] proteinemia are more prone to ASCVD than those without an elevation of plasma Lp [a] (Seed et al. 1990). In our lipid clinic at the University of Chicago, we have observed several cases of premature ASCVD with elevated plasma levels of Lp [a] associated with low levels of HDL. Exogenous factors such as cigarette smoking, diets rich in saturated fats, etc., should also be considered together with disease states such as hypertension, diabetes, renal and liver disorders leading to dyslipoproteinemias. Similarly, Lp [a] may be considered a precipitating factor in patients with a family history of fibrinolytic disorders. Overall, Lp [a] must be viewed as an important genetic determinant of ASCVD depending on the severity of the hyperLp [a] proteinemia and the presence of other risk factors, either atherogenic, thrombogenic or both.

Diagnostic Considerations

Ideally, adults and children alike should know their plasma Lp [a] levels along with the values of plasma total cholesterol, LDL and HDL cholesterol. In practice, however, considering the lack of standardized techniques for Lp [a] and the current unavailability of these techniques in clinical chemistry laboratories, it would be best to limit Lp [a] assays to subjects with a personal and/or family history of ASCVD, particularly in those with normocholesterolemia and premature ASCVD. The clinical importance of defining apo(a) phenotypes remains to be unequivocally established also in view of the unsettled nature of the assay. One of the most immediate goals should be to define the comparative pathogenicity of single versus double band apo(a) phenotypes.

Approaches to Correct High Plasma Levels of Lp [a] (Scanu and Fless 1990; Scanu 1990)

The correction of high plasma levels of CE-Lp [a] has proven to be a difficult task both by dietary means or by a combination of diets and hypolipidemic agents. Of the hypolipidemic agents used, only nicotinic acid or niacin in high dosage (4–5 g daily) have been shown to reduce plasma Lp [a] levels in the range of 20%–30% (Carlson et al. 1989). In those studies, however, not all subjects responded to the treatment and those who did respond still exhibited "unsafe" plasma levels of Lp [a]. The long-term side effects of niacin were also not evaluated and one has to question whether a pharmaceutical intervention is justified at this time. In this respect, we must point out that drugs of the statin

group have been found to be ineffective in modifying the plasma levels of
CE-Lp[a] (Brewer 1990). The correction of high plasma levels of TG-Lp[a]
invites a more optimistic outlook in that hypertriglyceridemic states are readily
manageable either by diets or a combination of diets and drugs. In this respect,
n-3 fatty acids and niacin should prove useful.

Acknowledgements

The original studies by the author and his associates were supported by
NIH-HNLBI Program Project grant no. 18577. Ms Sue Hutchison provided
invaluable help in preparing this manuscript.

References

Berg K (1990) Lp(a) lipoprotein: an overview. In: Scanu AM (ed) Lipoprotein(a). Academic Press,
    San Diego, pp 1–23
Brewer HB (1990) Effectiveness of diet and drugs in the treatment of patients with elevated Lp(a)
    levels. In: Scanu AM (ed) Lipoprotein(a). Academic Press, San Diego, pp 211–220
Carlson LA, Mansten A, Asplund A (1989) Pronounced lowering of serum levels of lipoprotein
    Lp(a) in hyperlipidaemic subjects treated with nictotinic acid. J Intern Med 226:271–276
Cushing GL, Gaubatz JW, Nava ML, Burdick BJ, Bocan TMA, Guyton JR, Weilbaecher D,
    DeBakey ME, Lawrie GM, Morrisett JD (1989) Quantitation and localization of apolipo-
    protein(a) and B in coronary artery bypass vein grafts resected at re-operation. Arteriosclerosis
    9:593–603
Fless GM, Pfaffinger DJ, Eisenbart ID, Scanu AM (1990) Solubility, immunochemical and
    lipoprotein binding properties of apoB$_{100}$-apo(a), the protein moiety of lipoprotein (a). J Lipid
    Res 31:909–918
Haberland ME, Fless GM, Scanu AM, Fogelman AM (1989) Modification of Lp(a) by
    malondialdehyde leads to avid uptake by human monocyte–macrophages. Arteriosclerosis 9:700a
Loscalzo J (1990) Lipoprotien(a): a unique risk factor for atherothrombotic disease. Arteriosclerosis
    10:672–679
McLean JW, Tomlinson JE, Kuang W, Eaton DL, Chen EY, Fless GM, Scanu AM, Lawn RM
    (1987) cDNA sequence of human apolipoprotein(a) is homologous to plasminogen. Nature
    330:132–137
Rath MS, Niendorf A, Reblin T, Dietel M, Krebber JH, Beisiegel U (1989) Detection and
    quantification of lipoprotein(a) in the arterial wall of 107 coronary bypass patients.
    Arteriosclerosis 9:570–592
Scanu AM (1990) Lipoprotein(a): A genetically determined cardiovascular pathogen in search of a
    function. J Lab Clin Med 116:142–146
Scanu AM, Fless GM (1990) Lipoprotein(a): heterogeneity and biological relevance. J Clin Invest
    85:1709–1715
Seed M, Hoppichler F, Reaveley D, McCarthey S, Thompson GR, Boerwinkle E, Utermann G
    (1990) Relation of serum lipoprotein(a) concentration and apolipoprotein(a) phenotype to
    coronary heart disease in patients with familial hypercholesterolemia. N Engl J Med
    322:1494–1496
Snyder ML, Fless GM, Polacek D, Scanu AM (1990) Human monocyte derived macrophages
    degrade Lp(a) differently than LDL. Arteriosclerosis 10:908a
Sorrentino M, Vielhauer C, Fless GM, Scanu AM, Feldman T (1990) Paradoxical difference in
    plasma lipoprotein(a) levels in white patients with two and three vessel coronary artery disease.
    Arteriosclerosis 10:829a
Utermann G (1989) The mysteries of lipoprotein(a). Science 246:904–910

Chapter 7

Modified Lipoproteins and Atherogenesis

J. A. Berliner, M. Territo, A. Amdalibi, M. Navab, F. Liao,
S. Cushing, S. Imes, J. Kim, B. Van Lenten, A. J. Lusis and
A. M. Fogelman

Introduction

A number of studies have shown that oxidized lipoproteins are present in the
artery wall and that the amount of oxidized lipid is increased in the
atherosclerotic plaque (Palinski et al. 1989; Haberland et al. 1988; Ylä-Herttuala
et al. 1989; Clevidence et al. 1983). Low-density lipoprotein (LDL) isolated
from the vessel wall has been shown to contain oxidized lipids and protein
(Ylä-Herttuala et al. 1989; Clevidence et al. 1983) and monoclonal antibodies to
malonidialdehyde lysine and hydroxynonenal show localization of this product
in lesions of the artery wall (Haberland et al. 1988; Palinski et al. 1989). These
oxidized lipoproteins have been suggested to play an important role in
atherogenesis from the pre-fatty streak stage through the definitive fibrous
plaque. Evidence of this role comes from studies showing that probucol and
other antioxidants can inhibit plaque formation in cholesterol-fed animals
(Carew et al. 1987; Kita et al. 1987).

The mechanisms by which oxidized lipoproteins can accelerate the
atherogenic process have been explored in in vitro studies by a number of groups
(Berliner et al. 1986; Quinn et al. 1985, 1987; Berliner et al. 1990; Rajavasisth et
al. 1990; Cushing et al. 1990; Hessler et al. 1983). Highly oxidized lipoproteins
have been shown to be toxic for dividing smooth muscle cells and fibroblasts
(Hessler et al. 1983), to chemotactically attract monocytes at high concentrations and to be capable of causing cholesterol loading of macrophages (Quinn et
al. 1985, 1987). These lipoproteins contain 20–50 nm of thiobarbituric acid
reactive substances (TBARS) per mg cholesterol and large amounts of lysine
modification of apolipoprotein B (apo B) allowing the particles to be taken up

by the scavenger receptor (Quinn et al. 1985, 1987). Alternatively they form aggregates which are taken up by phagocytosis.

The cells of the vessel wall have been shown to be capable of oxidizing lipoproteins in vitro (Morel et al. 1984; Cathcart et al. 1985; Parthasarathy et al. 1985, 1986; Hiramatsu et al. 1987; Steinbrecher et al. 1984); both superoxides and lipoxygenase products have been shown to be implicated in oxidation. Macrophages phagocytosing particles such as lipid droplets are an especially rich source of these oxidants.

Our laboratory has been examining the role of minimally modified LDL (MM-LDL) in processes that would cause fatty streak formation (Berliner et al. 1986, 1990; Rajavasisth et al. 1990; Cushing et al. 1990). In MM-LDL there is a low level of lipid oxidation (2–5 nm of TBARS/mg cholesterol as compared to 1–2 nm in native LDL); however, there is little oxidation of the protein, and therefore MM-LDL is taken up by the native LDL receptor. Minimally modified LDL is produced by storage of LDL at 4°C for three months to one year or by mild iron or copper oxidation. There are only low levels of lipopolysaccharide present in these preparations – less than 5 pg/µg of LDL protein. We have examined the effects of MM-LDL on chemotactic factor production, monocyte binding and production of colony-stimulating factors. In addition, we have documented the formation of MM-LDL from native LDL and have identified molecules that block oxidation and that inhibit the action of MM-LDL.

Methods

Rabbit aortic (REC), human aortic (HAEC) and human umbilical vein (HUV) endothelial cells have been employed at passages 7–14 for REC and 2–6 for HAEC and HUV. Human aortic smooth muscle cells at passages 5–7, the monocytic cell lines THP-1 and U937, and the granulocytic cell line HL60, have also been used. Freshly isolated human monocytes (using the modified Ricaldi method) and freshly isolated T-cells were used in some studies.

Binding of leukocytic cells to endothelial monolayers and chemotaxis studies were performed as previously described (Berliner et al. 1986). Measurements of colony-stimulating factors (CSF) levels were made using cultures of mouse bone marrow (Rajavasisth et al. 1990). For in vivo studies BALB/c mice were injected through a tail vein and tissues harvested after 5 hours. Nine mice were used for each experiment.

Results

Treatment of REC or HUV for 4 hours with MM-LDL caused an increase in the binding of monocytes but not neutrophils or T-cells to the endothelial monolayer. REC, HAEC and human aortic smooth muscle cells (HASMC) were incubated for 24 hours with various concentrations of native MM-LDL from 0.1 to 20 µg/ml and the amount of chemotactic activity for monocytes or neturophils in the medium was determined. In the presence of MM-LDL there was a dose-dependent increase in chemotactic activity for monocytes but not neutrophils. This activity was as much as tenfold over the basal amount produced in these cells. Unconditioned medium with up to 20 µg/ml of MM-LDL was not chemotactic. The increased chemotactic activity in the medium of MM-LDL treated cells was shown to be due entirely to the chemotactic peptide MCP-1. When conditioned medium was treated with antibody to MCP-1 (obtained from Dr A. Valente) all of the chemotactic activity for monocytes was abolished. There was a dose-dependent increase in the message for MCP-1 as determined by Northern blot analysis in both endothelial and smooth muscle cells treated with MM-LDL, and the amount of the increase paralleled the increase in chemotactic activity. After 4 hours of treatment of REC or HAEC with MM-LDL there was also an increase in the production of macrophage CSF and granulocyte macrophage CSF as assayed in a colony-forming assay using mouse bone marrow. This increase was paralleled by an increase in message for these colony-stimulating factors.

We have also examined the effect of MM-LDL in vivo. BALB/c mice were injected with phosphate buffered saline or MM-LDL (100 µg/ml blood) and the effects on leukocyte binding examined. After 5 hours there was a visible increase in binding of leukocytes to aortic orifices of intercostals (Fig. 7.1). There was about a twofold increase in binding per orifice when aortas from nine mice injected with PBS were compared to nine mice injected with MM-LDL.

Highly variable effects of cholesterol feeding have been observed in rabbits and other species. Some aortas from rabbits fed a high-cholesterol diet for two

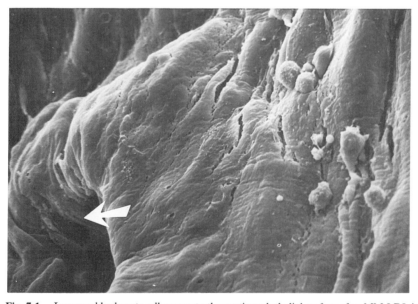

Fig. 7.1.   Increased leukocyte adherence to the aortic endothelial surface after MM-LDL injection. Four hours after the injection of MM-LDL, to give a concentration of 100 µg/ml MM-LDL in the blood, the aorta was perfusion fixed and examined by scanning electron microscopy (arrow indicates orifice) 3000×

months and exhibiting similar levels of LDL, HDL and beta very low density lipoprotein (BVLDL) exhibited essentially no lipid-filled lesions, whereas in others as much as one-third of the aorta was filled with lesion. We reasoned that in these resistant animals a protective molecule(s) might be present that inhibited the effects of oxidized LDL. In in vitro studies we have shown evidence for differential protection. HUV and some strains of REC are resistant to the toxic effects of high levels of MM-LDL, whereas other strains of REC are sensitive to these effects. Evidence has also been obtained that an induced protein may be responsible for this protection. When HUV are treated with MM-LDL in the presence of cycloheximide they become sensitive to its toxic effects and exposure of sensitive REC to low levels of MM-LDL induces protection against its toxicity (Berliner et al. 1990). Potential candidates for the protective proteins are currently being studied.

Discussion

The present in vitro and in vivo studies have shown that relatively minimal oxidation of LDL can cause profound changes in artery wall cells that may contribute to the formation of the fatty streak: induction of monocyte binding molecules, chemotactic factors and growth factors. Vessel wall cells (endothelium and smooth muscle cells) can produce this minimally oxidized LDL, making oxidation a plausible initiation mechanism in vivo. It might be expected that areas of the artery wall where LDL accumulates such as areas of predilection would be especially susceptible to LDL oxidation and fatty streak formation. We have also shown that protective cell proteins in addition to plasma lipoprotein levels may account for differences in susceptibility to atherosclerosis in different individuals.

Acknowledgements

This research was supported by NIH grants HL30568, HL26890, HL42550, RR865 and ESO 3466 and by the Laubisch Fund.

References

Berliner JA, Territo MC, Almada L, Carter A, Shafonsky E, Fogelman AM (1986) Monocyte chemotactic factor produced by large vessel endothelial cells in vitro. Arteriosclerosis 6:254–258
Berliner JA, Territo MC, Sevanian A, Ramin S, Kim JA, Bamshad B, Esterson M, Fogelman AM (1990) Minimally modified LDL stimulates monocyte endothelial interactions. J Clin Invest 85:1260–1266
Carew TE, Schwenke DC, Steinberg D (1987) An antiatherogenic effect of probucol unrelated to its hypocholesterolemic effect: evidence that antioxidants in vivo can selectively inhibit low density lipoprotein degradation in macrophage-rich fatty streaks slowing the progression of atherosclerosis in the WHHL rabbit. Proc Natl Acad Sci USA 84:7725–7729

Cathcart MK, Morel DW, Chisolm GM III (1985) Monocytes and neutrophils oxidize low density lipoproteins making it cytotoxic. J Leukocyte Biol 38:341–350

Clevidence BA, Morton RE, West G, Dusek DM, Hoff JF (1983) Cholesterol esterification in macrophages: stimulation by lipoproteins containing apo B isolated from human aortas. Arteriosclerosis 4:196–207

Cushing SD, Berliner JA, Valente AJ, Territo MC, Navab M, Parhami F, Gerrity R, Schwartz CJ, Fogelman AM (1990) Minimally modified LDL induces monocyte chemotactic proteins I in human endothelial and smooth muscle cells. Proc Natl Acad Sci USA 87:5134–5138

Haberland ME, Fong D, Cheng L (1988) Malondialdehyde-altered protein occurs in atheroma of Watanabe heritable hyperlipidemic rabbits. Science 241:215–218

Hessler JR, Morel DW, Lewis LJ, Chisolm GM (1983) Lipoprotein oxidation and lipoprotein-induced cytotoxicity. Arteriosclerosis 3:215–222

Hiramatsu K, Rosen H, Heinecke JW, Wolfbauer G, Chait A (1987) Superoxide initiates oxidation of low density lipoprotein by human monocytes. Arteriosclerosis 7:55–60

Kita T, Nagano Y, Yokode M, Ishii K, Kume N, Ooshima A, Yoshida H, Kawai C (1987) Probucol prevents the progression of atherosclerosis in Watanabe heritable hyperlipidemic rabbit, an animal model for familial hypercholesterolemia. Proc Natl Acad Sci USA 84:5928–5931

Morel DW, DiCorleto PE, Chisolm GM (1984) Endothelial and smooth muscle cells alter low density lipoprotein in vitro by free radical oxidation. Arteriosclerosis 4:357–364

Palinski W, Rosenfeld ME, Ylä-Herttuala S, Gurtner GC, Socher SS, Butler SW, Parthasarathy S, Carew TE, Steinberg D, Witztum JL (1989) Low density lipoprotein undergoes oxidative modification in vivo. Proc Natl Acad Sci USA 86:1372–1376

Parthasarathy S, Steinbrecher UP, Barnett J, Witztum JL, Steinberg D (1985) Essential role of phospholipase A_2 activity in endothelial cell-induced modification of low density lipoprotein. Proc Natl Acad Sci USA 82:3000–3004

Parthasarathy S, Printz DJ, Boyd D, Joy L, Steinberg D (1986) Macrophage oxidation of low density lipoprotein generates a modified form recognized by the scavenger receptor. Arteriosclerosis 6:505–510

Quinn MT, Parthasarathy S, Steinberg D (1985) Endothelial cell-derived chemotactic activity for mouse peritoneal macrophages and the effects of modified forms of low density lipoprotein. Proc Natl Acad Sci USA 82:5949–5953

Quinn MT, Parthasarathy S, Fong LG, Steinberg D (1987) Oxidatively modified low density lipoproteins: a potential role in recruitment and retention of monocyte/macrophages during atherogenesis. Proc Natl Acad Sci USA 84:2995–2998

Rajavasisth TB, Andalibi A, Territo MC, Berliner JA, Navab M, Fogelman AM, Lusis AJ (1990) Modified LDL induce endothelial cell expression of granulocyte and macrophage colony stimulation factors. Nature 344:254–257

Steinbrecher UP, Parthasarathy S, Leake DS, Witztum JL, Steinberg D (1984) Modification of low density lipoprotein by endothelial cells involves lipid peroxidation and degradation of low density lipoprotein phospholipids. Proc Natl Acad Sci USA 83:3883–3887

Ylä-Herttuala S, Palinski W, Resenfeld ME, Parthasarathy S, Carew TE, Butler S, Witztum JL, Steinberg D (1989) Evidence for the presence of oxidatively modified low density lipoprotein in atherosclerotic lesions of rabbit and man. J Clin Invest 84:1086–1095

Chapter 8

The Molecular Biology of Apolipoprotein B

J. Scott

Introduction

An elevated level of circulating cholesterol is a most important risk factor for atherosclerosis. Cholesterol is carried in the plasma as low-density lipoprotein (LDL). LDL is an oily droplet, containing a core of hydrophobic cholesteryl esters and a surface monolayer of phospholipid and free cholesterol at the surface. The particle is stabilized and made soluble in the water of the blood by the presence at its surface of a huge protein apolipoprotein (apo) B (Scott 1989a).

Apo B contributes to atherogenesis in three ways (Scott 1989a). First and most importantly, LDL which has undergone free radical damage is taken up by macrophages, the progenitors of the atherogenic foam cell (Steinberg et al. 1989). Second, LDL binds directly to negatively charged proteoglycans like heparin in the artery wall, and so accumulates in the arterial intima. Third, apo B is covalently associated with another protein, apo(a), in particles called Lp[a] (Scott 1989b). Increased levels of Lp[a] are another major risk factor of atherosclerosis. Lp[a] is homologous to plasminogen, the zymogen of the fibrinolytic enzyme plasmin. It blocks plasminogen binding to vascular endothelial cells, and thereby inhibits fibrinolysis and promotes surface coagulation. Thus apo B contributes to both atherogenesis and thrombogenesis.

Hepatic Apo B Metabolism

Apo B-100 is the hepatic form of apo B (Scott 1989a) (Fig. 8.1). Apo B-100 is secreted from the liver as a triglyceride-rich particle called very-low-density

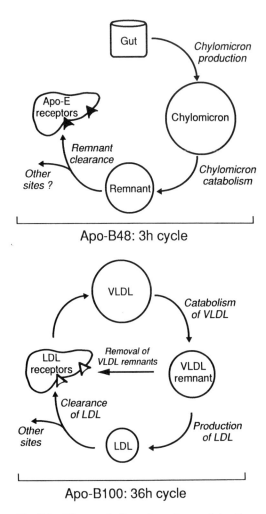

Fig. 8.1. The metabolism of apo B-containing lipoproteins.

lipoprotein (VLDL). VLDL circulates to peripheral capillaries, where it is transiently bound to heparin and undergoes an interaction with lipoprotein lipase. This enzyme hydrolyzes triglyceride and facilitates the uptake of liberated fatty acid by fat cells. The VLDL remnant circulates back to the liver, where it can undergo two distinct fates. It may be cleared into the liver by interaction with the LDL receptor, or interact with hepatic lipase which hydrolyzes the remainder of the triglyceride. All of the small apolipoproteins are lost from the surface of the particle as it shrinks in size. The remnant of this metabolism is highly enriched in cholesterol. It contains apo B as its sole protein component, and is called LDL. LDL delivers cholesterol to all tissues of the body by the LDL receptor pathway. It is the balance between production of apo B by the liver, and clearance by the LDL receptor pathway which determines blood cholesterol concentrations.

Structure and Function of Hepatic APO B-100

Apo B-100 as a Ligand for the LDL Receptor and a Heparin-Binding Protein

It is known from chemical modification and other studies that the regions of apo B that bind to the LDL receptor and to heparin are enriched in the positively charged arginine and lysine residues. Eight regions of apo B enriched in basic residues exist (Fig. 8.2). The two basic regions, residues 3147–3157 (A) and 3357–3367 (B), with the highest density of positive charge reside on either side of amino acid residue 3249, which is one of the two thrombin cleavage sites in apo B (Fig. 8.3). These regions have also been shown by Weisgraber and colleagues to have the highest affinity for heparin (Scott 1989a). The sequence of the basic peptide B is remarkably similar to that of the LDL receptor-binding

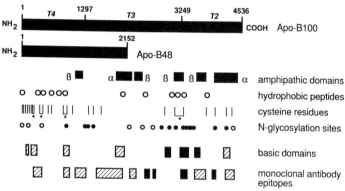

Fig. 8.2. The structure of apo B-100. Thrombin cleavage fragments T2, T3 and T4 are shown. Amphipathic α-helical domains are designated α, and proline-rich β-sheets β. Cysteine residues marked by an asterisk are known to be cross-linked. *N*-glycosylation sites shown as filled circles are known to be utilized. Two basic domains in the region of apo B-100 that binds to the LDL receptor are shown as filled boxes A and B. Monoclonal antibody epitopes for antibodies that block binding to the LDL receptor are shown as filled boxes.

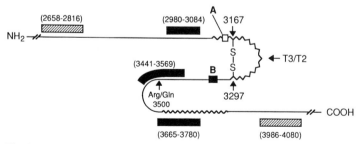

Fig. 8.3. The LDL receptor binding domain of apo B-100. The thrombin cleavage site between fragments T3 and T2 is shown. Proline-rich amphipathic β-sheets are shown as zigzag lines. Basic peptides A and B are shown. Monoclonal antibody epitopes for antibodies that completely block the binding of apo B-100 to the LDL receptor are shown as filled boxes. Antibodies that partially block binding to the LDL receptor are shown as hatched boxes. The mutation of Arg to Gln at position 3500 is shown.

Table 8.1.   Basic peptides of Apo B-100 and Apo E

Apo B A	^{3147}-*Lys*-Ala-Gln-Tyr-*Lys*-*Lys*-Asn-*Lys*-*His*-*Arg*-*His*-^{3157}
Apo B B	^{3357}-Thr-Thr-*Arg*-Leu-Thr-*Arg*-*Lys*-*Arg*-Gly-Leu-*Lys*-^{3367}
Apo E	^{140}-*His*-Leu-*Arg*-*Lys*-Leu-*Arg*-*Lys*-*Arg*-Leu-Leu-*Arg*-^{150}
LDL receptor (consensus)	Cys-Asp-X-X-X-Asp-Cys-X-Asp-Cly-Ser-Asp-Glu

domain of apo E, the other ligand for the LDL receptor (Table 8.1) (Knott et al. 1986). The basic residues in apo B and apo E both bind to acid residues in the LDL receptor.

To further define the LDL receptor binding domain of apo B, we have used a series of monoclonal antibodies which have been mapped to apo B fusion proteins expressed in *E. coli* and to tryptic fragments of apo B (Knott et al. 1986; Milne et al. 1989; Pease et al. 1990). Thus, we have been able to demonstrate that the LDL receptor-binding domain spans approximately 1200 amino acids, which encompasses the basic peptides A and B.

Our current view of the structure of the LDL receptor binding domain of apo B-100 is shown in Fig. 8.3. The basic peptides A and B reside at the junction of thrombin cleavage site at residue 3249. Comparative sequencing studies on the chick and six mammalian species have shown that the basic peptide B, which is homologous to the receptor-binding domain of apo E, is highly conserved in primary and secondary structure (Law and Scott 1990). The basic peptide A is not conserved. Evidence from a variety of sources indicates that this region is tightly folded. First, receptor-blocking monoclonal antibodies bind to sequences which are considerably distant in the primary structure, but compete with each other for binding, indicating close proximity (Milne et al. 1989; Pease et al. 1990). Second, tryptic mapping indicates the presence of the disulfide bridge between residues 3167 and 3297. Finally, highly conserved proline-rich structures are likely to form amphipathic β-pleated sheets, which bind this region of apo B to the lipoprotein particle.

Further corroboration for this region being involved in LDL receptor binding has come from the discovery by Innerarity and colleagues of a mutation causing substitution of arginine codon 3500 with glutamine (Soria et al. 1989). This variant was discovered in a patient with severe hypercholesterolemia and produces defective binding of apo B to the fibroblast LDL receptor.

Apo B as a Lipid-Binding Protein

The lipid-binding properties of apo B serve the functions of lipid transport and lipoprotein assembly (Scott 1989a) (Fig. 8.2). It is known from a variety of studies that two types of amphipathic structure mediate the lipid-binding functions of the apolipoproteins. In the small, typical apolipoproteins such as apo E, tandemly repeated amphipathic α-helices are involved in lipid binding. These structures describe 3.7 residues per α-helical turn and have hydrophobic residues on one face of the helix and hydrophylic on the other. Amphipathic β-strands are also involved in lipid binding by apo B, but are not found in the smaller apolipoproteins. Both of these types of structure are extensively distributed in apo B so that there are numerous anchor points throughout the length of apo B which bind it to the lipid of the lipoprotein particle.

In addition, amphipathic structures are clustered in several regions of apo B (Fig. 8.2). Two extensive amphipathic α-helical regions are found: one in the middle of apo B-100 and one in a carboxy terminal. Proline-rich amphipathic β-sheets are found in four regions of the molecule. Two of these proline-rich regions are in the LDL receptor-binding domain, with the others towards the amino terminus and in the mid-portion of apo B. In addition, a number of short hydrophobic peptides have been identified; none of these except for the signal peptide at the amino terminal would be long enough to cross the plasma membrane. Finally, apo B has been shown to be linked covalently to fatty acid at cysteine residues (Hoeg et al. 1988) and the amino terminal of apo B is particularly cysteine rich. All of the different lipid-binding structures in apo B have a feature in common, that is that they would all serve to bind apo B to the monolayer at the surface of the lipoprotein particle, and to the inner leaflet of the membranes within the cell that are involved in lipoprotein assembly and secretion. None of the lipid-binding domains is long enough to cross a cellular membrane.

Apo B-100 and Lipoprotein Assembly

The current knowledge about lipoprotein assembly and secretion is summarized in Fig. 8.4. It is known from a number of studies that apo B takes some 10 minutes to translate on the ribosome (Scott 1989a). During this process it is tightly associated with the membranes of the endoplasmic reticulum (ER) and it is at this site that it undergoes a variety of post-translational modifications. Significant among these is covalent linkage with apo (a) to form Lp[a] (Scott 1989b). Relatively rapidly after translation is known to be complete, apo B can be detected loosely bound or free in the lumen of the smooth ER. At this stage it has received lipids from their sites of synthesis in the outer leaflet of the membranes of the ER, from which they must be translocated to the inner leaflet where apo B resides. Apo B must disassociate from the membranes of the ER as a nascent lipoprotein and become free in the lumen. From the smooth ER apo B passes to the Golgi apparatus, where further lipid is added before the lipoprotein particle is secreted. None of these processes are understood. It is known from

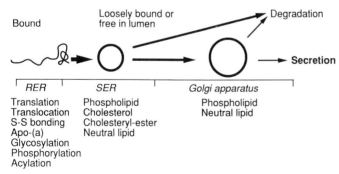

Fig. 8.4. Apo B-containing lipoprotein assembly. The sites of post-translational modification and lipid addition to the lipoprotein particle are shown.

the work of Olofsson and Davies and from our own laboratory that a substantial component of the apo B that is synthesized is degraded within the cell and not secreted (Scott 1989a). Modulation of the amount of apo B secreted or degraded is a major factor in regulating apo B-containing lipoprotein secretion. The transcription of the apo B gene is not influenced by metabolic factors (Pullinger et al. 1989).

Two disorders shed light on the process of lipoprotein assembly and secretion. These are abetalipoproteinemia and hypobetalipoproteinemia. Both are characterized by failure to produce apo B-containing lipoproteins in the intestine and the liver. As a consequence, affected individuals develop fat and fat-soluble vitamin malabsorption. The latter leads to spinocerebellar degeneration and retinitis pigmentosum. The difference between these conditions is that abetalipoproteinemia is an autosomal recessive disorder and hypobetalipo-proteinemia is a co-dominant disorder.

We have shown in two affected families with abetalipoproteinemia that RFLP haplotypes at the apo B locus do not cosegregate with this disorder (Talmud et al. 1988). For an autosomal recessive disorder the affected sibling should receive the same alleles at the apo B gene if the disease is caused by a defect of the apo B gene. We therefore concluded that abetalipoproteinemia is not a defect of the apo B gene but of a gene which facilitates apo B production. The defect presumably resides in the lipoprotein assembly and secretion pathways, because apo B can be demonstrated in the cells of patients with this disorder.

We have also had the opportunity to study individuals with hypobetalipo-proteinemia, and in this disorder have identified two mutations of the apo B gene which predict truncated forms of apo B (Collins et al. 1988). These have been designated apo B-39 and apo B-26 on the centile system because they represent these percentages of apo B-100. When we came to examine lipoprotein fractions from the patient predicted to express apo B-39, a truncated variant was found in the VLDL fraction. For the patient with the apo B-26 mutation, no truncated form of apo B was found in any of the lipoprotein fractions. We therefore argued that apo B-26 was either not secreted from the cell or that the lipoprotein particle was unstable.

To examine this issue, we prepared a number of truncated constructs of apo B including apo B-39, B-26, B-23, B-17 and B-13 named on the centile system, and expressed these in the hepatoblastoma cell line HepG2. Lipoprotein and non-lipoprotein fractions were prepared and immunoprecipitated. Cells were transfected with apo B-39 and one of the other truncated variants. Only a trace of B-17 and no B-13 was present in the lipoprotein fractions. The other truncated forms were present in lipoproteins. In the infranatant fraction there was no apo B-39, but there were substantial quantities of apo B-26, B-23, B-17 and B-13. We therefore concluded that the region delineated by apo B-13 and B-39 contains structures essential for the formation and secretion of stable lipoprotein particles. It is likely that there are lipid-binding structures throughout this domain and that together they are necessary for lipoprotein integrity and for the unique characteristic of apo B among the apolipoproteins, that it does not exchange between lipoprotein particles.

To summarize thus far: apo B has a cysteine-rich amino terminal, essential features for lipoprotein assembly and integrity are contained between apo B-13 and B-39, and the LDL receptor-binding domain resides in the carboxy terminal and spans some 1200 amino acids.

Intestinal Apo B-48 Metabolism

Intestinal apo B is called apo B-48 because it is 48% of the size of apo B-100 on SDS–polyacrylamide gels. It is essential for triglyceride-rich chylomicron assembly in the intestine. Chylomicrons undergo the same fate as hepatic VLDL, and deliver fat to peripheral tissues (Fig. 12.1). The chylomicron remnant circulates back to the liver, where it is cleared by a lipoprotein receptor, distinct from the LDL receptor that binds apo E and not apo B. This is the chylomicron remnant receptor.

Structure and Biosynthesis of Apo B-48

Monoclonal antibody-binding studies have indicated that apo B-48 represents the amino terminal half of apo B-100. A priori, three well-known biological mechanisms could be responsible for the production of apo B-48. Either there could be two apo B genes, one producing apo B-48 and the other apo B-100, or there could be differential splicing of apo B-48 and apo B-100 exons from a single RNA transcript, or differential proteolytic processing of apo B-100.

Examination of intestinal RNA from humans and rabbits with 5' and 3' probes showed a 14.5-kb apo B-100 mRNA which hybridized with both probes, and a smaller 7-kb species which hybridized with the 5' probe only. This suggested the possibility that the short mRNA produced apo B-48. We therefore examined intestinal cDNA clones in the region where it was predicted that apo B-48 would terminate. Two differences between the apo B-100 cDNA structure and the intestinal cDNA structures were identified. First, there were polyadenylated cDNAs providing the molecular basis for the small mRNA. Second, in clones representing both the large and small mRNAs examined there was a C to T change at nucleotide 6666. This C to T change predicted that amino acid codon 2153 for glutamine in the hepatic mRNA would become UAA in the intestinal mRNA and would encode a stop translation codon at the exact position where protein sequencing subsequently showed that apo B-48 ended (Chen et al. 1987; Powell et al. 1987; Higuchi et al. 1988).

This was surprising to us, because we had previously determined the intron/exon organization of the apo B gene (Fig. 8.5) (Scott 1989a). The position where apo B-48 ended was in the middle of the huge exon 26, and we had found by extensive sequence analysis that there was no basis for differential splicing. In further experiments we were able to demonstrate that there was no other copy of the apo B gene (Powell et al. 1987). Moreover we examined the intestinal genomic DNA and found no evidence for the stop translation codon in the genome. These and other studies led us to the unprecedented conclusion that apo B-100 and apo B-48 were both products of the same gene and that in liver the mRNA produced led in a straightforward fashion to the production of apo B-100 (Fig. 12.5). However, in the intestine the apo B RNA underwent a co- or post-transcriptional modification, in which nucleotide 6666 was modifed from C to U, and this led to termination of apo B-48 translation at amino acid residue 2052. In addition it was speculated that this mechanism somehow revealed cryptic polyadenylation signals, which led to the production of a short message

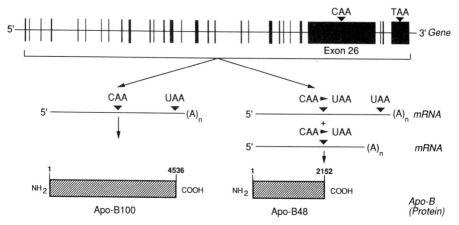

Fig. 8.5. The intron–exon organization of the apo B gene and schematic representation of the mechanism for production of apo B-100 and apo B-48 mRNAs.

in the intestine, but not in the liver. The latter has recently been confirmed (Boström et al. 1989).

To understand this unusual problem we have developed an in vitro RNA modification system, and we have used this to define the sequences involved in the modification reaction, and are currently using it to attempt the purification of the activity that modifies apo B mRNA.

To facilitate these studies we developed a primer extension analysis using reverse transcriptase and dideoxy GTP, which could cause termination in human transcripts at position 6666 if there was a C in that position or at 6655, the next C if there was a T at position 6666. Fortuitously, the primer extension in the mouse and rat terminated at 6661 if there was a T at position 6666 (Davies et al. 1989; Driscoll et al. 1989). The conversion assay was performed on synthetic apo B-100 mRNA with S100 cytoplasm extracts taken from the rat hepatoma McArdle 7777, which produces apo B-48, and was assayed by the primer extension analysis. We were able to demonstrate that in the presence of a high concentration of EDTA we could effect conversion in vitro. That conversion had occurred was confirmed by cloning and sequencing. Conversion was evidently not dependent on the presence of divalent cations and did not require exogenous or endogenous nucleotide triphosphates. Proteinase K treatment destroyed the activity, which was therefore protein dependent (Driscoll et al. 1989). The activity was not destroyed by micrococcal nuclease. We therefore believe the activity is likely to be an enzyme which is capable of recognizing and modifying apo B mRNA.

To delineate the minimum sequence requirements for editing in vitro, a variety of constructs between 2.3 kb and 26 nucleotides were made (Driscoll et al. 1989). Sequences between 2.3 kb and 55 nucleotides underwent editing, but the construct of 26 nucleotides was not edited. Consideration of the secondary structure in this region of apo B in comparative studies between human, rabbit, rat and mouse predicted that a short stem-loop is present in this region and that the edited nucleotide always occurred in an 8-bp loop (Davies et al. 1989). The 26-bp construct, which is asymmetrical, disrupted this stem-loop. We are

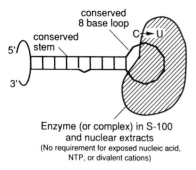

Fig. 8.6.   Model for the mechanism of apo B mRNA editing.

currently involved in a series of mutagenesis studies to test whether this secondary structure is important for editing.

Our model for RNA editing is therefore as follows. It seems likely that a conserved stem-loop is involved in editing. The edited base occurs in an 8-base loop (Fig. 8.6). This part of the model is being tested. An enzyme is likely to be involved in editing. The enzyme has no requirement for divalent cations or for ATP. The enzyme is probably similar to those involved in the modification of tRNA. Once this enzyme has been purified we can address the mechanism of C to U change.

Regulation of RNA Editing

In rodents, the situation is different from that in humans; both apo B-100 and apo B-48 are produced by the liver. It is known that the level of production of the two proteins can be modulated metabolically. A most dramatic metabolic modification occurs in rats with altered thyroid hormone status. Thus, hypothyroid animals produce mainly B-100 and hyperthyroid animals produce mainly B-48 from the liver. We have now shown that this modulation acts by regulating the level of the stop codon in apo B mRNA (Davidson et al. 1988).

We also sought to examine whether apo B mRNA could be edited in mouse tissues other than in the liver and intestine. We found that mice expressed high levels of endogenous apo B mRNA in tissues as diverse as the brain and testis, and that in both of these tissues there was dramatic editing of apo B mRNA. This study suggests that, in the mouse at least, the two forms of apo B must have a role in local lipid homeostasis, and suggests the possibility that editing enzyme may act on other mRNAs in different tissues.

Conclusions

What is the imperative for nature to have evolved this unusual mechanism for producing two forms of apo B from the same gene? Lipoproteins derived from the gut and containing apo B-48 circulate back to the liver with a cycle time of approximately 3 hours (Fig. 12.1). In contrast, the residence time of LDL apo B-100 in the circulation is 36 hours. The need here seems to be to get triglyceride derived from the gut to the periphery and to get some of the intestinal cholesterol and fat-soluble vitamins rapidly back to the liver. The structural difference between these proteins determines their very different behavior, and it may well be the presence or absence of the LDL receptor binding domain, and the need to deliver intestinally derived lipid rapidly to the liver, which necessitated the invention of apo B-48.

References

Boström K, Lauer SJ, Poksay KS, Garcia Z, Taylor J, Innerarity TL (1989) Apolipoprotein B48 RNA editing in chimeric apolipoprotein EB mRNA. J Biol Chem 264:15701–15708

Chen S-H, Habib G, Yang CY et al. (1987) Apolipoprotein B-48 is the product of a messenger RNA with an organ-specific in-frame stop codon. Science 238:363–366

Collins DR, Knott TJ, Pease RJ et al. (1988) Truncated variants of apolipoprotein B cause hypobetalipoproteinaemia. Nucl Acids Res 16:8361–8375

Davidson NO, Powell LM, Wallis SC, Scott J (1988) Thyroid hormone modulates the introduction of a stop codon in rat liver apolipoprotein B messenger RNA. J Biol Chem 263:13482–13485

Davies MS, Wallis SC, Driscoll DM et al. (1989) Sequence requirements for apolipoprotein B RNA editing in transfected rat hepatoma cells. J Biol Chem 264:13395–13398

Driscoll DM, Wynne JK, Wallis SC, Scott J (1989) An in vitro system for the editing of apolipoprotein B mRNA. Cell 58:519–525

Higuchi K, Hospattankar AV, Law SW, Meglin N, Cortright J, Brewer HB Jr (1988) Human apolipoprotein B (apoB) mRNA: identification of two distinct apoB mRNAs, an mRNA with the apoB-100 sequence and an apoB mRNA containing a premature in-frame translational stop codon, in both liver and intestine. Proc Natl Acad Sci USA 85:1772–1776

Hoeg JM, Meng MS, Ronan R, Demosky SJ Jr, Fairwell T, Brewer HB Jr (1988) Apolipoprotein B synthesized by Hep G2 cells undergoes fatty acid acylation. J Lipid Res 29:1215–1220

Knott TJ, Pease RJ, Powell LM et al. (1986) Complete protein sequence and identification of structural domains of human apolipoprotein B. Nature 323:734–738

Law A, Scott J (1990) A cross-species comparison of the apolipoprotein B domain that binds to the LDL receptor. J Lipid Res 31:1109–1120

Milne R, Theolis R Jr, Maurice R et al. (1989) The use of monoclonal antibodies to localize the low density lipoprotein receptor-binding domain of apolipoprotein B. J Biol Chem 264:19754–19760

Pease RJ, Milne RW, Jessup WK et al. (1990) Use of bacterial expression cloning to localize the epitopes for a series of monoclonal antibodies against apolipoprotein B100. J Biol Chem 265:553–568

Powell LM, Wallis SC, Pease RJ, Edwards YH, Knott TJ, Scott J (1987) A novel form of tissue-specific RNA processing produces apolipoprotein-B48 in intestine. Cell 50:831–840

Pullinger CR, North JD, Teng B-B, Rifici VA, de Brito AER, Scott J (1989) The apolipoprotein B gene is constitutively expressed in HepG2 cells: regulation of secretion by oleic acid, albumin, and insulin, and measurement of the mRNA half-life. J Lipid Res 30:1065–1077

Scott J (1989a) The molecular and cell biology of apolipoprotein-B. Mol Biol Med 6:65–80

Scott J (1989b) Thrombogenesis linked to atherogenesis at last? Nature 341:22–23

Soria LF, Ludwig EH, Clarke HRG, Vega GL, Grundy SM, McCarthy BJ (1989) Association between a specific apolipoprotein B mutation and familial defective apolipoprotein B-100. Proc Natl Acad Sci USA 86:587–591

Steinberg D, Parthasarathy S, Carew TE, Khoo JC, Witztum JL (1989) Beyond cholesterol: modifications of low-density lipoprotein that increase its atherogenicity. N Engl J Med 320:915–924

Talmud PJ, Lloyd JK, Muller DPR, Collins DR, Scott J, Humphries S (1988) Genetic evidence from two families that the apoliopoprotein B gene is not involved in abetalipoproteinemia. J Clin Invest 82:1803–1806

Chapter 9

Structure and Evolution of the Apolipoprotein and Lipase Gene Families

L. Chan, W. Hide, Yau-Wen Yang and Wen-Hsiung Li

Introduction

The plasma lipoproteins are the transport vehicles that deliver lipids from the liver and intestine to the peripheral tissues for disposal, and in turn, they capture some of the peripherally derived lipids and return them to the liver, where they are metabolized. Lipoprotein lipase (LPL) and hepatic triglyceride lipase (HL) are enzymes located on the vascular endothelial surface that modify, through lipid hydrolysis, the circulating lipoprotein particles. On the surface of these particles are the apolipoproteins, the protein constitutents which are essential for the structure and stability of the different lipoprotein classes. In addition to their structural role, the apolipoproteins also serve other diverse functions, e.g. as ligands for specific cell surface receptors (apo E and apo B-100) and enzyme activators (apo C-II and apo A-I).

We have been interested in the evolution of the apolipoproteins and the vascular lipases for a number of years. Our studies indicate that we are dealing with two structurally unrelated gene families: the apolipoprotein multigene family and the lipase gene family. By examining the rate of evolution of various subdomains in each protein group, we have learned much concerning the structure–function relationships in these proteins. We shall review our analyses pertaining to each family, with special emphasis on the most recent experimental data.

Methods

The method of Li et al. (1985) was used to estimate nucleotide substitution rates between homologous coding regions. Nucleotide sites are classified according to degrees of synonymity, and substitutions are estimated in terms of the number of substitutions per synonymous site (K_S) and per non-synonymous site (K_A).

In order to compare conserved structure/function relationships among the lipases, we developed a conservation index (CI) for each gene: pairs of genes from within each family were aligned and given a consensus score for conserved residues at each site. A score of 1 was given if residues were the same at a particular position. A score of 0 was given if residues differed at the same position. A window size of nine residues was found to give the most useful resolution for comparison of conservation. The average identity (conservation index) for each nine-residue window was determined, and the index value was plotted against the median residue position.

The Soluble Apolipoprotein Multigene Family

About ten years ago, Barker and Dayhoff (1977), Fitch (1977) and McLahlan (1977) independently observed that apo A-I contains multiple repeats of 22 amino acids (22-mer), each of which is a tandem array of two 11-mers. The repeat unit of 22-mer has been suggested to be a structural element that builds an amphipathic α-helix (Fukushima et al. 1979; Kaiser and Kezdy 1983). Recently, Nakagawa et al. (1985) proposed that a 44-mer composed of two 22-mers punctuated in the middle by a helix breaker, Pro, rather than the 22-mer itself (Sparrow and Gotto 1982), is the paradigm of lipid-binding domains of apo A-I. They showed that the 44-mer mimics more closely the surface properties and conformation of apo A-I than does the 22-mer. They speculated that the centrally located Pro residue that breaks the α-helix actually keeps the hydrophobic faces in phase and the resulting concavity of the 44-mer is suited for the absorption of the peptide to the highly curved surface of human plasma HDL_3 (radius 40–50 Å).

The existence of a 22-mer periodicity has subsequently been found in other apolipoproteins, including apo A-II, A-IV, C-II, C-III and E, and an 11-mer has been found in apo C-I (Luo et al. 1986; Boguski et al. 1984, 1985, 1986; Das et al. 1985; Elshourbagy et al. 1986; Karathanasis et al. 1986).

Using the junction between exon 3 and intron 3 as a reference point, Luo et al. (1986) obtained an alignment of the genes for the above proteins for the 33 codons upstream from the junction. In this block of 33 codons, the homologies between genes are, in the majority of cases, about 40% or higher at the nucleotide level, and 15% or higher at the amino acid level (Luo et al. 1986). Luo et al. (1986) suggested that the three segments in the common block arose from triplication of an 11-mer (11 amino acids); these segments have been designated A, B or C. The similarity between segments A and B is higher than that between segments B and C. Therefore, it appears that segments B and C were first derived from a duplication of 11 codons and later segment A was derived from a duplication of segment B.

When the junction between intron 3 and exon 4 is used as a reference point, a pattern of internal repeats in exon 4 also becomes evident for each of the genes considered (Luo et al. 1986). The fundamental unit of repeat in exon 4 is not an 11-mer, but a 22-mer that is made up of two 11-mers. The main reason for this is that most 11-mers are more similar to the 11-mer one unit removed than to the adjacent 11-mer (Luo et al. 1986; Fitch 1977; Karathanasis et al. 1983). This basic structure of 22-amino acid units connected to one another by a proline residue is highly reminiscent of the 22- and 44-amino acid structures described as the basic lipid-binding domains by Fukushima et al. (1979) and Nakagawa et al. (1985). Indeed, some of the repeats represent lipid-binding amphipathic helices identified in the individual proteins (Sparrow and Gotto 1982). Luo et al. (1986) have designated the three repeats of 11 codons in exon 3 the first three repeats in each gene and the first 22 codons in exon 4 the fourth repeat, e.g. A-I-4.

Evolution of Apo A-I and Apo E

Apo A-I is an activator of the enzyme lecithin:cholesterol acyltransferase and is the major protein in high-density lipoproteins. It was the first apolipoprotein noted to have internal repeats (Barker and Dayhoff 1977; Fitch 1977; McLahlan 1977). Apo E is a ligand for the low-density lipoprotein receptor and a specific apo E receptor (Mahley 1988). We have recently cloned cow apo A-I (O'hUigin et al. 1990) and apo E cDNAs, and we will present our results on the molecular evolution of these two proteins.

Pairwise comparisons of rat with other mammalian (cow, dog, human, macaca and rabbit) apo A-I sequences consistently yield higher substitution rates (five comparisons; mean $K_A = 0.24$, $K_S = 0.65$) than do other mammalian pairwise combinations (nine comparisons; mean $K_A = 0.11$, $K_S = 0.30$, with the human–monkey comparison excluded). High substitution rates are commonly found in rodent genes (Li et al. 1987), the average rate being 1.5 times the mammalian average. Luo et al. (1989) have shown that the substitution rate of apo A-I and apo E in the rat lineage is three to four times greater than that in the human lineage. We confirm this observation here by using three additional mammalian lineages. Apparently the rat apo A-I is atypical of mammalian apo A-I genes (see below).

To identify conserved and therefore potentially functionally important regions of apo A-I, nucleotide-substitution patterns over short segments were examined. A 90-nucleotide rolling average of K_A is shown in Fig. 9.1. This figure includes

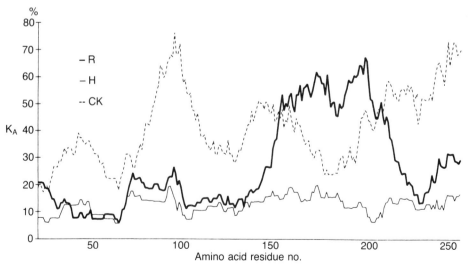

Fig. 9.1. K_A, measured in 30-codon rolling average across the aligned apo A-I sequences of cow and rat (R), cow and human (H), and cow and chicken (CK).

Table 9.1. K_A and K_S in divergent and conserved regions of the rat apo A-I gene

Comparison	K_A		K_S		N^b
	Conserved	Diverged[a]	Conserved	Diverged[a]	
Mammals to mammals (human to macaca excluded)	0.104	0.124	0.317	0.307	9
Rat to mammals	0.189	0.517	0.650	0.603	5
Chicken to mammals	0.394	0.356	0.978	0.757	5

Note: mammalian sequences from cow, dog, human, macaca and rabbit were compared with sequences from either rat or chicken, and substitution rates were estimated. These estimates were weighted by the number of sites used in each comparison, and a mean estimate was made. The human–macaca comparison was excluded from the intramammalian comparison, since these species belong to the same order.

[a] Refers to codons 150–203 of the rat prepropeptide.

[b] Number of pairwise comparisons.

cow–rat, cow–human and cow–chicken comparisons. Substitutions are rather evenly distributed in the cow–human comparison but are unevenly distributed in the cow–rat comparison. A particularly high substitution rate defines a region of the rat gene, 162 nt in length and flanked by short deletions, encoding amino acid residues 150–203 of the rat protein. The mean K_A in the cow–rat comparison increases from 0.18 outside this region to 0.51 within this region (Fig. 9.1). This localized high substitution rate is also seen in other rat–mammalian comparisons (not shown). Table 9.1 shows that, in the divergent region, the mean K_A between rat and other mammals undergoes an almost threefold increase and exceeds the mean K_A of mammal–bird comparisons. K_S and K_A are almost equal (0.60 vs. 0.52) within the divergent region for rat–mammal comparisons. Within the conserved region of apo A-I, the mean K_A for rat–mammal comparisons is 1.8 times greater than that among the mammals, close to the 1.9-fold higher rate found by Li et al. (1987) for the average of rodent genes. In contrast, the rate is 4.2 times greater in the divergent region of the rat gene.

The evolutionary history of apo E is likely to be significantly different from that of apo A-I, because apo E contains a well-defined receptor binding domain which is expected to be well conserved through evolution. Table 9.2 shows the K_S and K_A values seen in pairwise comparisons of eight mammalian apo E sequences. A detailed phylogenetic analysis using these sequences and sequence data from other genes (Li et al. 1990) strongly suggests that the bovine lineage is more closely related to the dog lineage than to any of the other lineages in Table 9.2, and that there is no close relationship between rodents and lagomorphs (rabbits) as suggested by some paleontologists (e.g. Simpson 1945; Novacek 1982). The large K_S and K_A values between the guinea pig and the myomorph (mice and rats) apo E genes suggest that the guinea pig and the myomorph rodents are not closely related to each other, though the large values could be partly due to higher rates of nucleotide substitution in rodents (Li et al. 1990).

The K_A value between cow and dog is larger than those between either of them and human, monkey or rabbit. Therefore, if the bovine and dog lineages

Table 9.2. Number of substitutions per synonymous site (K_S) and per non-synonymous site (K_A) for apo E

	Cow	Dog	Human	Baboon	Rabbit	Rat	Mouse	Guinea pig
A. K_S (below diagonal) and SE (above diagonal)								
Cow		0.048	0.053	0.054	0.059	0.108	0.107	0.073
Dog	0.297		0.048	0.055	0.051	0.094	0.092	0.067
Human	0.371	0.309		0.026	0.049	0.082	0.082	0.055
Baboon	0.373	0.385	0.117		0.053	0.086	0.085	0.059
Rabbit	0.414	0.325	0.329	0.369		0.091	0.087	0.062
Rat	0.786	0.693	0.618	0.644	0.707		0.033	0.106
Mouse	0.794	0.689	0.616	0.638	0.676	0.168		0.107
Guinea pig	0.522	0.467	0.381	0.427	0.439	0.783	0.775	
B. K_A (below diagonal) and SE (above diagonal)								
Cow		0.020	0.018	0.018	0.020	0.023	0.023	0.020
Dog	0.206		0.018	0.018	0.019	0.022	0.021	0.019
Human	0.168	0.166		0.007	0.016	0.019	0.018	0.017
Baboon	0.173	0.159	0.028		0.016	0.018	0.018	0.016
Rabbit	0.202	0.188	0.137	0.138		0.021	0.021	0.019
Rat	0.248	0.225	0.182	0.171	0.209		0.010	0.019
Mouse	0.244	0.218	0.170	0.164	0.211	0.057		0.019
Guinea pig	0.207	0.180	0.149	0.136	0.177	0.176	0.178	

Note: binomial names for the species used are as follows: Huma = *Homo sapiens*, rat = *Rattus norvegicus*, dog = *Canis familiaris*, cow = *Bos taurus*, rabbit = *Oryctolagus cuniculus*.

are indeed closer to each other than to the primate and rabbit lineages, then the apo E sequence has evolved faster in the bovine and dog lineages than in the latter lineages. Using the human, monkey, rabbit, rat, mouse and guinea pig sequences as references, we estimate that K_A values in the bovine and dog lineages are respectively 0.112 and 0.094 since the divergence of the two lineages. This computation suggests that the apo E sequence has evolved 20% faster in the bovine lineage than in the dog lineage.

A comparison of the K_A values in Table 9.2 with those in Table 1 of O'hUigin et al. (1990) reveals that apo E evolves somewhat faster than apo A-I except in the rodent lineage. For example, the K_A value between human and rabbit is 0.137 for apo E and 0.119 for apo A-I.

To identify conserved and therefore potentially functionally important regions of apo E, we have aligned all available apo E sequences (Fig. 9.2). In the alignment the symbol + on the top of a site signifies that all the sequences have the same amino acid at that site, while the symbol # signifies that all the sequences have "similar" amino acids at that site. By "similar" amino acids we mean amino acids that are hydrophobic (M, V, L, I, F, Y and W), that are basic (R and K), that are acidic (D and E), or that are neither proline (which causes a disruption in secondary structure) nor in any of the preceding groups (see Fitch 1977). Apo E, like other soluble apolipoproteins, is almost completely made up of internal repeats, i.e. a common 33-codon block (which is located immediately upstream from the third exon/intron junction and consists of three repeats of 11-mer) and repeats E-4 to E-14 (see Luo et al. 1986). In Fig. 9.2 the common block and each of the other repeats are separated by a space and are overlined.

Apo E is an important determinant in the interaction between apo E-containing lipoproteins and cell-surface receptors (Mahley and Innerarity 1983). Studies using monoclonal antibodies, natural mutants and site-specific mutants produced in vitro localized the receptor-binding region to the vicinity of residues 140–150 and have thus far identified at least eight specific residues (nos 136, 140, 142, 143, 145, 150 and 158 of the human mature peptide) as crucial residues involved in receptor binding (Wardell et al. 1987; Lalazar et al. 1988). Further, the α-helical conformation in this region also appears to be essential, since substitution of Pro for either Leu 144 or Ala 152 will interfere with binding activity. All of these sites except residues 136, 137 and 138 are located within repeat E-8 (residues 139–160). It is seen that residues 140, 142, 143, 146, 150 and 158 have been conserved in all sequences and that site 144 is occupied by only hydrophobic residues. Further, residues 136 and 152 have also been conserved in all but one of the sequences. Thus, all the crucial residues have been rather well conserved among the sequences.

Human apo E is polymorphic with three major alleles. The three isoforms corresponding to these alleles differ in the amino acid residues in positions 112 and 158. Apo E-4 has arginine and apo E-2 cysteine in both positions. Apo E-3, the commonest isoform, has Cys 112 and Arg 158. The receptor-binding activity of apo E-2 is impaired due to the Arg 158 $\rightarrow$ Cys substitution, and the inheritance of apo E-2 is a major predisposing factor in Type III hyperlipoproteinemia. It is interesting that residue 112 is either Cys or Arg in eight different species examined, whereas residue 158 is exclusively Arg in all cases (Fig. 9.2), suggesting that an Arg 158 $\rightarrow$ Cys substitution in the apo E of other species may also be deleterious to its receptor-binding function. From this analysis, we conclude that the ancestral apo E could have been an analog of either apo E-3 or apo E-4 but not apo E-2.

The 33-codon block is very well conserved in apo E among all the sequences studied. In fact, 23 out of the 33 sites are identical among all the sequences, five other sites are occupied by "similar" amino acids, and among the five sites that are not signified by + or # three are conserved in all but one sequence. Thus, it is the best conserved region in apo E and this fact suggests that it serves an important function.

Repeat E-14 is also well conserved among the sequences. In fact, it is better conserved than the receptor-binding region, though not as well conserved as the 33-codon block. Thus, this region may be functionally or structurally important.

It is obvious that the internal repeats are much more conservative than the flanking N- and C-terminal regions. However, among the internal repeats, those near the N- and C-terminal ends are more conservative than those in the middle part of the protein. In this connection it is interesting to note the experimental observation that in aqueous solution apo E contains two independently folded domains: a relatively unstable self-associating carboxyl-terminal domain (residues 225–299) rich in amphipathic helices, and a more stable amino-terminal domain (residues 20–165) that resembles a soluble globular protein in structure (Wetterau et al. 1988; Aggerbeck et al. 1988). The two domains are connected by an exposed peptide segment or hinge region that appears to have random coil structure and is highly susceptible to proteolysis. The differential conservation observed in Fig. 9.2 supports the thesis that the two structural domains require more stringent sequence conservation for their function than does the hinge region.

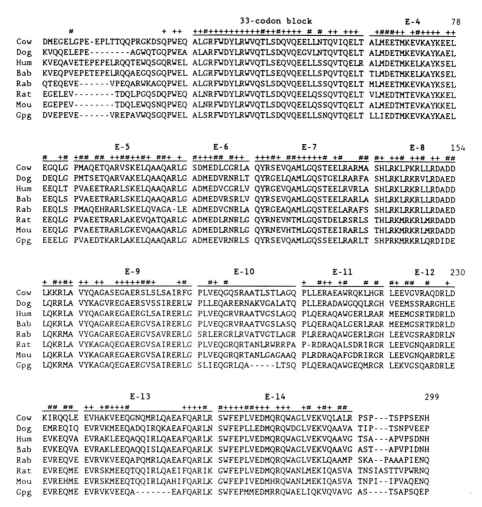

Fig. 9.2. Sequence alignment of mammalian apo E amino acid sequences. Hum, human; Bab, baboon; Rab, rabbit; Mou, mouse; Gpg, guinea pig. The symbols on top are, +, identical residues in all sequences; #, similar residues as defined in the text.

Structure and Evolution of the Lipase Gene Family

LPL and HL show considerable homology with pancreatic lipase (PL), and the three enzymes are members of a gene family (Datta et al. 1988; Semenkovich et al. 1989) distinct from the apolipoprotein gene family. Furthermore, these lipases show considerable similarity to one region of the *Drosophila* yolk proteins, YP1, YP2 and YP3 (Ameis et al. 1990).

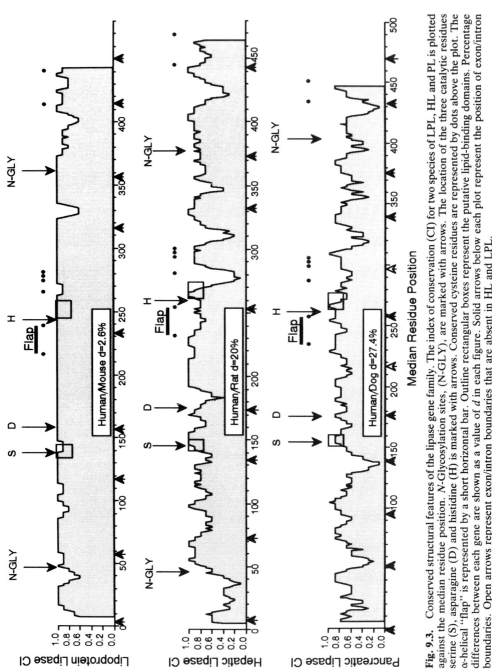

Median Residue Position

Fig. 9.3. Conserved structural features of the lipase gene family. The index of conservation (CI) for two species of LPL, HL and PL is plotted against the median residue position. *N*-Glycosylation sites, (N-GLY), are marked with arrows. The location of the three catalytic residues serine (S), asparagine (D) and histidine (H) is marked with arrows. Conserved cysteine residues are represented by dots above the plot. The α-helical "flap" is represented by a short horizontal bar. Outline rectangular boxes represent the putative lipid-binding domains. Percentage differences between each gene are shown as a value of *d* in each figure. Solid arrows below each plot represent the position of exon/intron boundaries. Open arrows represent exon/intron boundaries that are absent in HL and LPL.

Conservation and Structure

Using a moving-window conserved residue analysis, we constructed an index of conservation (CI). We compared the position of putative functional structures to their CI values (Fig. 9.3). We found a correlation between a high CI and the position of conserved functional structures.

The relative amount of variation between each pair of genes depends on two factors: first, the general degree of conservation exhibited by the type of gene, and second, the genetic distance between each gene pair. For instance, comparison between the LPL of human and mouse reveals a high degree of identity (Fig. 9.3). The gene in question is highly conserved from species to species. Human and mouse LPL sequences demonstrate a corrected Dayhoff distance of 2.6% (Dayhoff 1978).

As the LPLs are highly conserved, we found it informative to perform an additional CI analysis on all the available sequence data from five different species, as shown in Fig. 9.4. Within the LPL gene family there is a marked degree of conservation at regions sharing known functions. CI values of 0.8 or higher are demonstrated at putative lipid-binding sites, catalytic residues, possible structurally important cysteine residues and N-glycosylation sites. Chicken LPL has a C-terminal putative N-glycosylation site that differs in position from the mammalian N-glycosylation sites.

Catalytic Residues and Lipid-Binding Regions

The most highly conserved feature in all lipases is the segment containing the putative lipid-binding site (Datta et al. 1988) and the "catalytic" serine residue (Fig. 9.3; also see Figure 2 in Datta et al. 1988). This segment is very similar in vertebrate lingual/gastric lipase, prokaryotic lipases and vertebrate lecithin:cholesterol acyltransferase. The lipid-binding site is characterized by a region of hydrophobic residues. Within this region is a consensus sequence G–x–S–x–G similar to the consensus sequence of active site serines in serine esterases and serine proteases (Brenner 1988).

X-ray crystallography of human PL has shown that the catalytic serine residue is located in the N-terminal domain at the C-terminal edge of a doubly wound parallel β-sheet, and is part of a Ser–Asp–His triad (Winkler et al. 1990). The crystalline structure of human PL suggests that substantial conformational changes occur before it can bind substrate in this postulated active site. There is a surface loop between disulfide-bridged cysteine residues (237–261) that covers the active site with a short one-turn α-helix. This "flap" has to be repositioned before the site can become accessible to substrate. Reference to sequence alignments (see Datta et al. 1988) reveals that the amino acid sequence of the flap region is not highly conserved between the different lipases. However, there are consistently aligned cysteine residues in all the lipases bounding the flap region. Reference to Figs 9.3 and 9.4 reveals that the region has varying degrees of conservation within each of the genes. It is thus likely that the functional site of lipases is covered in each gene by a loop of amino acid residues serving only as structural components of a short α-helical flap which may be quite flexible in primary structure. Predicted α-helical moment data for the region in each of the lipases show that despite the highly dissimilar amino acid sequences there is a

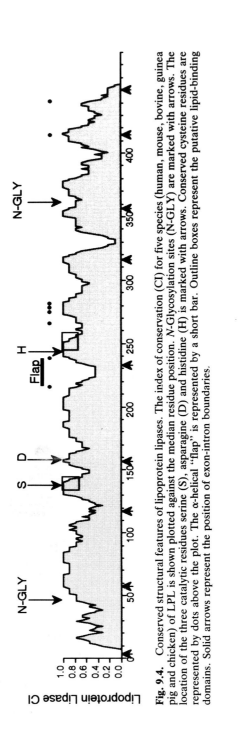

Fig. 9.4. Conserved structural features of lipoprotein lipases. The index of conservation (CI) for five species (human, mouse, bovine, guinea pig and chicken) of LPL is shown plotted against the median residue position. *N*-Glycosylation sites (N-GLY) are marked with arrows. The location of the three catalytic residues serine (S), asparagine (D) and histidine (H) is marked with arrows. Conserved cysteine residues are represented by dots above the plot. The α-helical "flap" is represented by a short bar. Outline boxes represent the putative lipid-binding domains. Solid arrows represent the position of exon-intron boundaries.

good probability that an α-helix is present in the "flap" region in all of the lipases (data not shown).

The lipid-binding site lies in a region where the conservation index is 1.0 for the LPLs and the HLs and 0.8 for the PLs (see Fig. 9.3). The region is bounded on either side by residues that are less well conserved in HL and PL (CI < 0.7 for HL and PL). Each amino acid of the three-part catalytic triad lies in a segment showing a high degree of conservation (CI > 0.8 for PL and CI = 1.0 for HL and LPL).

There is a second predicted hydrophobic lipid-binding region (Bengtsson-Olivecrona et al. 1986; Datta et al. 1988) which also lies in a highly conserved region (CI = 1.0 for LPL and HL, CI = 0.8 for PL). The high CI for the predicted region is suggestive of a conserved function for the second putative lipid-binding site.

N-Glycosylation Sites

Two putative N-glycosylation sites are conserved in the mammalian lipoprotein and hepatic lipases (the second site differs in position in chicken LPL). The regions in which they lie are not as highly conserved as the lipid-binding domains but show conservation at or above 0.8 (Fig. 9.3). The position is conserved in hepatic and lipoprotein lipases for both sites. The positions of the putative N-glycosylation sites in pancreatic lipase do not align in the same positions as the other lipases. It is interesting that mammalian LPL requires proper N-glycosylation for activity (Semenkovich et al. 1990) whereas the PL of at least one species, the horse, is totally devoid of carbohydrate (Chaillan et al. 1990).

Cysteine Residues

There are eight cysteine residues that align to the same place in all members of the lipase family. The conserved position of these residues probably reflects their role in formation of disulfide bridges required for maintenance of enzyme structure and function. Six of the eight residues are found in the central very highly conserved region of the family. One of the residues, at median residue position 268 in LPL, lies in a region that has a CI ≤ 0.8 for the equivalent position in LPL, HL and PL. In LPL, a seventh cysteine residue at the C-terminal region of the gene (median residue position 415 in LPL) is located at or near a peak of conservation. This suggests that the cysteine residue at this position is likely to have an important structural function. The eighth cysteine residue is entirely conserved in all LPLs yet lies 11 residues from the end of the gene in all cases except for the longer chicken LPL gene, where the cysteine lies 27 residues from the end of the gene. The last cysteine residue does not lie in a region of high conservation in any of the lipases and is less likely to have a structural role.

HL has a more varied CI profile than LPL for the distribution of cysteine residues. This reflects the greater structural variation within the HL and the greater genetic distance of the two genes being compared. The final cysteine residue at median residue position 470 is very near the end of the protein, has a low CI and so may not have a structural role. Because the window of comparison between proteins is nine residues in size, it is not possible to perform adequate comparisons of conservation when within nine residues of the end of proteins being compared.

Exon/Intron Structure

The exon/intron structure of human LPL, HL and canine PL was compared against the conservation profile as shown in Fig. 9.3. Exon/intron boundary positions are not correlated with degree of conservation. Human LPL is composed of ten exons. Human HL has nine exons and dog PL has 13 exons. Exon sizes and boundary positions are very similar in HL and LPL (Cai et al. 1989; Ameis et al. 1990), although LPL has an extra exon comprised exclusively of the 3' untranslated region of the gene. The similarity in exon/intron boundary distribution between LPL and HL suggests they have diverged more recently than PL.

PL has a distinctly different organization of exon/intron boundaries with respect to the other lipases. PL has extra introns, "splitting" the exon organization. For example, exons 4 and 5 of PL are analogous to exon 3 of HL and LPL and exons 7 and 8 of PL are analogous to exon 5 of HL and LPL.

HL and LPL may have lost several introns after divergence from a common ancestor to the lipases. As it is likely that PL has the greatest similarity to the common ancestral gene, it is therefore likely that the ancestral lipase from which the lipase gene family arose contained the same number of, or more, introns than modern PL.

Domains and Exons

Pancreatic lipase has a marked degree of similarity in conservation profile with HL between median residues 140–350 (Fig 9.3). In particular, the lipid-binding domains are at or near peaks of high CI values, and cysteine residues and the residues of the catalytic triad are at conserved peaks. The conserved region is represented by exons 4, 5 and 6 in LPL and HL. The same region is represented by exons 6, 7, 8, 9 and 10 in PL. It appears that HL and LPL have lost two introns from the conserved region.

Exon shuffling has been proposed as a mechanism for the evolution of multidomain genes and has been inferred in the LDL receptor (Südhof et al. 1985) and serine proteases (Rogers 1985). Exon shuffling may have played a role in the evolution of the lipases. For example, both putative lipid-binding domains of the lipase gene family are contained within distinct exons. The equivalent lipid-binding domain located in exon 4 of HL and LPL and exon 6 of PL is bounded on either side by an exon/intron boundary. A major domain border has been suggested around residue 228 in human LPL (Bengtsson-Olivecrona et al. 1986). An exon/intron boundary lies just to the right of this residue.

Catalytic Function

The catalytic function of the lipases may have arisen after the lipid-binding function. The region of similarity with the *Drosophila* vitellogenins contains a putative lipid-binding region. Although highly similar to lipase sequences, this region in the vitellogenins does not have a conserved "catalytic" serine residue.

The catalytic serine residue is replaced by an asparagine in YP1 and a glycine in YP2, which may mean that YP is not catalytically active. It has been suggested that the serine residue in lipases is not an active-site residue (Verger 1984; Persson et al. 1989). However, that is not consistent with results of site-directed mutagenesis experiments on human LPL (Faustinella and Chan, unpublished results) and structural evidence from human PL (Winkler et al. 1990).

If the YP genes contain domains that share ancestry with the lipases, and exon shuffling was one of the processes which resulted in the structure of the modern lipases, then the catalytic function of lipases has arisen after the lipid-binding function. The putative lipid-binding domains are contained within discrete exons, but the three-part putative catalytic site is made up of residues in discrete exons (which are contiguous in HL and LPL, but non-contiguous in PL (see Fig. 9.3)). Thus the catalytic function could not have resulted directly from an exon shuffling event.

Concluding Remarks

We have analyzed the structure–function and evolutionary relationships of the apolipoprotein and lipase gene families. As additional data become available, we will be able to clarify some of the unexpected findings, e.g. the "unusually" high rate of nucleotide substitution in the guinea pig in both families. The importance of the various highly conserved domains in individual proteins will also become evident as their crystal structures become available. For example, for the lipase gene family, we understand at the molecular level the structure–function relationships of these enzymes much better than the apolipoproteins because of the availability of crystal structure of one of its members – PL.

Acknowledgement

The work reviewed in this chapter has been supported by grants from the National Institutes of Health HL-16512 (to L.C.) and GM-39927 (to W.-H.L.).

References

Aggerbeck LP, Wetterau KH, Weisgraber C-S, Wu C, Lindgren FT (1988) Human apolipoprotein E3 in aqueous solution. II. Properties of the amino- and carboxyl-terminal domains. J Biol Chem 263:6249–6258
Ameis D, Stahnke G, Kobayashi J (1990) Isolation and characterization of the human hepatic lipase gene. J Biol Chem 265:6652–6555
Barker WC, Dayhoff MO (1977) Evolution of lipoproteins deduced from protein sequence data. Comp Biochem Physiol 576:309–315
Bengtsson-Olivecrona G, Olivecrona T, Jörnvall H (1986) Lipoprotein lipases from cow, guinea-pig and man: structural characterization and identification of protease-sensitive internal regions. Eur J Biochem 161:281–288

Boguski MS, Elshourbagy N, Taylor JM, Gordon JI (1984) Rat apolipoprotein A-IV contains 13 tandem repetitions of a 22 amino acid segment with amphipathic helical potential. Proc Natl Acad Sci USA 81:5021–5025

Boguski MS, Elshourbagy N, Taylor JM, Gordon JI (1985) Comparative analysis of repeated sequences in rat apolipoproteins A-I, A-IV and E. Proc Natl Acad Sci USA 82:992–996

Boguski MS, Freeman M, Elshourbagy NA, Taylor JM, Gordon JI (1986) On computer-assisted analysis of biological sequences: proline puncutation, consensus sequences, and apolipoprotein repeats. J Lipid Res 27:1011–1034

Brenner S (1988) The molecular evolution of genes and proteins: a tale of two serines. Nature 334:528–530

Cai S-J, Wong DM, Chen S-H, Chan L (1989) Structure of the human hepatic triglyceride lipase gene. Biochemistry 28:8966–8971

Chaillan C, Abousalam K, Kerfelec B et al. (1990) Horse pancreatic lipase and colipase. CEC-GBF International Workshop, Braunschweig, Germany

Das HK, McPherson J, Bruns GAP, Karathanasis SK, Breslow JL (1985) Isolation, characterization and mapping to chromosome 19 of the human apolipoprotein E gene. J Biol Chem 260:6240–6247

Datta S, Luo C-C, Li W-H et al. (1988) Human hepatic lipase: cloned cDNA sequence, restriction fragment length polymorphisms, chromosomal localization and evolutionary relationships with lipoprotein lipase and pancreatic lipase. J Biol Chem 263:1107–1110

Dayhoff MO (1978) Atlas of protein sequence and structure (vol 5). National Biomedical Research Foundation, Silver Spring, Maryland

Elshourbagy NA, Walker DW, Boguski MS, Gordon JI, Taylor JM (1986) The nucleotide and derived amino acid sequence of human apolipoprotein A-IV mRNA and the close linkage of its gene to the genes of apolipoproteins A-I and C-III. J Biol Chem 261:1998–2002

Fitch WM (1977) Phylogenetics constrained by the cross-over process as illustrated by human hemoglobins and a thirteen cycle, eleven amino acid repeat in human apolipoprotein A-I. Genetics 86:623–644

Fukushima D, Kupferberg JP, Yokoyama S, Kroon DJ, Kaiser ET, Kezdy FJ (1979) A synthetic amphipathic helical docasapeptide with the surface properties of plasma apolipoprotein A-I. Am Chem Soc 101:3703–3704

Havel RJ, Kane JP, Kashyap ML (1973) Interchange of apolipoproteins between chylomicrons and high density lipoproteins during alimentary lipemia in man. J Clin Invest 52:32–38

Kaiser ET, Kezdy FJ (1983) Secondary structures of proteins and peptides in amphipathic environments (a review). Proc Natl Acad Sci USA 80:1137–1143

Karathanasis SK, Zannis VI, Breslow JL (1983) Isolation and characterization of the human apolipoprotein A-I gene. Proc Natl Acad Sci USA 80:6147–6151

Karathanasis SK, Yunis I, Zannis VI (1986) Structure, evolution, and tissue-specific synthesis of human apolipoprotein A-IV. Biochemistry 25:3962–3970

Komaromy MC, Schotz ME (1987) Cloning of rat hepatic lipase cDNA: evidence for a lipase gene family. Proc Natl Acad Sci USA 84:1526–1530

Lalazar A, Weisgraber KH, Rall SC Jr et al. (1988) Site specific mutagenesis of human apolipoprotein E: receptor binding activity of variants with single amino acid substitutions. J Biol Chem 263:3542–3545

Li W-H, Wu C-I, Luo, C-C (1985) A new method for estimating synonymous and nonsynonymous rates of nucleotide substitution considering the relative likelihood of nucleotide and codon changes. Mol Biol Evol 2:150–174

Li W-H, Tanimura M, Sharp PM (1987) An evaluation of the molecular clock hypothesis using mammalian DNA sequnces. J Mol Evol 25:330–342

Li W-H, Gouy M, Sharp PM, O'hUigin C, Yang Y-W (1990) Molecular phylogeny of rodentia, lagomorpha, primates, ardiodactyla, and carnivora and molecular clock. Proc Natl Acad Sci USA 87:6703–6707

Luo C-C, Li W-H, Moore MN, Chan L (1986) Structure and evolution of the apolipoprotein multigene family. J Mol Biol 187:325–340

Luo C-C, Li W-H, Chan L (1989) Structure and expression of dog apolipoprotein A-I, E, and C-I mRNAs: implications for the evolution and functional constraints of apolipoprotein structure. J Lipid Res 30:1735–1746

Mahley RW (1988) Apolipoprotein E: cholesterol transport protein with expanding role in cell biology. Science 240:522–530

Mahley RW, Innerarity TL (1983) Lipoprotein receptors and cholesterol homeostasis. Biochim Biophys Acta 737:197–222

McLahlan AD (1977) Repeated helical pattern in apolipoprotein A-I. Nature 267:465–466

Nakagawa SH, Lau HSH, Kezdy FJ, Kaiser ET (1985) The use of polymer-bound oximes for the synthesis of large peptides usable in segment condensation: synthesis of a 44 amino acid amphipathic peptide model of apolipoprotein A-I. J Am Chem Soc 107:7087–7092

Novacek MJ (1982) Information for molecular studies from anatomical and fossil evidence on higher eutherian phylogeny. In: Goodman M (ed) Macromolecular sequences in systematic and evolutionary biology. Plenum Press, New York, pp 3–41

O'hUigin CO, Chan L, Li W-H (1990) Cloning and sequencing of bovine apolipoprotein A-I cDNA and molecular evolution of apolipoproteins A-I and B-100. Mol Biol Evol 7:327–339

Persson B, Bengtsson-Olivecrona G, Enerbäck S, Olivecrona T, Jörnvall H (1989) Structural features of lipoprotein lipase. Eur J Biochem 179:39–45

Rogers J (1985) Exon shuffling and intron insertion in serine protease genes. Nature (Lond) 315:458–459

Semenkovich CF, Wims M, Noe L, Etienne J, Chan L (1989) Insulin regulation of lipoprotein lipase activity in 3T3-L1 adipocytes is mediated at posttranscriptional and posttranslational levels. J Biol Chem 264:9030–9038

Semenkovich CF, Luo C-C, Nakanishi MK, Chen S-H, Smith LC, Chan L (1990) In vitro expression and site-specific mutagenesis of the cloned human lipoprotein lipase gene. J Biol Chem 265:5429–5433

Simpson GG (1945) The principles of classification and a classification of mammals. Bull Am Museum Nat Hist 85:1–350

Sparrow JT, Gotto AM Jr (1982) Apolipoprotein/lipid interactions: studies with synthetic polypeptides. CRC Crit Rev Biochem 13:87–107

Südhof TC, Russel DW, Goldstein JL, Brown MS, Sanchez-Pescador R, Bell G (1985) Cassette of 8 exons shared by genes for LDL receptor and EGF precursor. Science 228:893–895

Verger R (1984) Lipases. Borgström B, Brockman HL (eds), Elsevier, Amsterdam, pp 83–150

Wardell MR, Brennan SO, James ED, Fraser R, Carrell RW (1987) Apolipoprotein E2-Christchurch (136 Arg → Ser): new variant of human apolipoprotein E in a patient with Type III hyperlipoproteinemia. J Clin Invest 80:483–490

Wetterau JR, Aggerbeck LP, Rall SC, Weisgraber KH (1988) Human apolipoprotein E3 in aqueous solution. I. Guidance for two structural domains. J Biol Chem 263:6240–6248

Winkler FK, D'Arcy A, Hunziker W (1990) Structure of human pancreatic lipase. Nature 343:771–774

Chapter 10

A Gln to Arg Substitution in the Adducin Family of Proteins is a Necessary but not Sufficient Factor for High Blood Pressure in Rats of the Milan Hypertensive Strain

F. E. Baralle

Introduction

Previous studies have shown that erythrocytes from the Milan hypertensive strain of rats (MHS) differ from the control normotensive strains (MNS). These differences are determined within the stem cells, are genetically associated with the development of hypertension, and are similar to those found between the tubular cells of the two strains. This alteration seems to be associated with abnormalities of membrane skeleton (Bianchi et al. 1985; Ferrari et al. 1986, 1987; Bianchi et al. 1990).

MHS rats immunized with ghosts or membrane skeleton extracts of MNS rats produced antibodies against a 105-kDa protein (Salardi et al. 1989) with many characteristics of human erythrocyte adducin (Gardner and Bennett 1986). Adducin is a 200-kDa membrane skeletal protein isolated from human erythrocyte ghost membranes. This protein is an α β heterodimer with subunits of $M_r = 103\,000$ (α) and $97\,000$ (β) subsequently isolated from bovine brain membrane (Bennett et al. 1988). Brain adducin contains two subunits of $M_r = 104\,000$ and $107\,000/109\,000$ immunologically related to erythrocyte adducin. Erythrocyte and brain adducin bind to calmodulin in a calcium-dependent manner and are phosphorylated by protein kinases-C. Furthermore adducin associates with spectrin–actin complexes and promotes association of additional spectrin molecules at the spectrin–actin junction. Isoforms of this protein have been detected in lung, kidney, testes and liver (Bennett et al. 1988). These features support the idea that adducin plays an important role in the cell membrane skeleton organization. Each adducin subunit contains two domains: a 39-kDa protease-resistant region, highly conserved between the α and β

subunits, and a 60–64-kDa protease-sensitive region which is unique to each subunit. The conserved domains are located at the amino-terminal region of adducin, and alone are unable to bind calmodulin, do not interact with spectrin and actin and lack the phosphorylation sites (Joshi and Bennett 1990). Similarity of protein sequence in some portions of the variable domains cannot be excluded.

Using the antibodies against the 105-kDa protein a cDNA clone from a mouse spleen expression library was isolated (Salardi et al. 1989) and used in the present study to isolate the rat adducin gene and cDNAs. Overlapping cDNA clones allow us to complete the sequence of an adducin-like protein of 63 kDa whose identity was corroborated by the homology of rat amino acid sequences with peptide sequences derived from human erythrocyte adducin. The cDNAs contained common sequences with at least four mRNAs in different tissues and the mHS and MNS rat gene pool were shown to present at least one amino acid polymorphism that might be associated to the blood pressure phenotypes.

Results

Isolation and Partial Identification of a Rat Adducin Genomic Clone

A rat adducin genomic clone was isolated using the mouse cDNA clone p3A1.1 (Salardi et al. 1989) as probe. The 5'-end region contains a short open reading frame of 183 bp highly homologous to the murine probe. An intron–exon junction canonic signal and an in-frame TGA stop codon followed by a non-coding region with a putative polyadenylation site, corresponding to the last exon of adducin gene, were found (Fig. 10.1B).

Isolation of Rat Spleen cDNA Clones

cDNA was synthesized from rat spleen poly(A^+) mRNA using the adducin-specific antisense OLIGO 3 oligonucleotide as primer. The cDNA was cloned and the recombinants carrying the longer adducin inserts (SNOC 31 and SNOC 33, Fig. 10.1C and D, respectively) were characterized by restriction enzyme mapping and DNA sequencing. This analysis revealed that SNOC 31 and SNOC 33 were overlapping and that SNOC 31 was 49 bp longer at the 5'-end. The 3'-ends of both clones ended with the OLIGO 3 oligonucleotide sequence.

In order to complete the mRNA sequence, a new cDNA library was constructed using the OLIGO 4 oligonucleotide as specific antisense primer corresponding to the 5'-end sequence of SNOC 31. The longer clone-isolated SNOC 431 was sequenced (Fig. 10.1E).

To confirm that our clones contained the 5'- and 3'-ends of adducin 63 mRNA the extremes were amplified and cloned using the "RACE protocol" (Frohman et al. 1988) The 5'-end RACE was carried out with GAT 7 oligonucleotide and the longer clone isolated (SHOGAT 7, Fig. 10.1F) contained ten additional bases 5' of SNOC 431. This was confirmed by primer extension experiments using the OLIGO 4 antisense oligonucleotide as primer for reverse transciption:

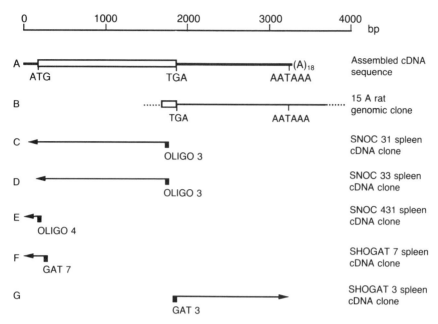

Fig. 10.1. Cloning strategy and organization of cDNA and genomic clones for adducin 63. (**A**) Assembled adducin 63 cDNA sequence; open box, open reading frame of 1863 nucleotides delimited by start and stop codon; solid line, untranslated region of adducin 63 cDNA. (**B**) Adducin rat genomic clone. Open box, last translated exon region; solid line, untranslated 3' region with the polyadenylation signal. (**C, D, E**) Overlapping cDNA clones derived from two specific cDNA libraries starting with OLIGO 3 and OLIGO 4 oligonucleotides. (**F, G**) cDNA clones generated by 5'-end and 3'-end PCR amplification according to the RACE protocol (Frohman et al. 1988).

one major product was found, extending ~150 nucleotides beyond the primer (data not shown).

The 3'-end RACE was carried out with the GAT 3 oligonucloetide and one of several positive clones isolated was sequenced (SHOGAT 3, Fig. 10.1G), confirming the presence of the poly(A) tail 18 nucleotides 3' of the AATAAA polyadenylation signal.

The Adducin 63 Sequence

The composite sequence of the overlapping cDNA clones (Fig. 10.2) revealed a 1683-bp open reading frame from the ATG initiation codon at position 187 resulting in a translation product of 63 kDa. The ATG initiation codon is preceded by three in frame stop codon at positions 172, 116, 70. The presence of the stop codon at position 172 was confirmed in the genomic DNA by polymerase chain reaction (PCR) amplification and sequencing (data not shown). It should be noted that all the independent cDNA clones and RACE-derived cDNA clones isolated from this region presented identical sequence (data not shown). The TGA stop codon at position 1870 is followed by a non-coding region of 1372 bp ended with a poly(A) tail.

```
     1 AGGCCACAGGCACTGCCTCCTTCATTTGTGGGAGTCTACAAGGATGGAGGATAGTTCCCACCCGGCTGTGTGAAGTGTGGCTGTGCCAGCCCTCCTAGTT
   101
     GTGAAGACCTGGCACTGAGGACAGAGACTGGACAGGCCTGGTTCATCCTACTGGGTTCCGGGAGTGGACTGAAGCCTTAGCAGGAGCATGAGTGAGGACACGGTCCCCGAGGCAGCCTCA
                                                                                   M  S  E  D  T  V  P  E  A  A  S
   221                                                                            1                             11
     CCGCCACCCTCTCAGGGGCAGCACTACTTTGACCGGTTCTCTGAGGATGACCCTGAATACTTGCGACTTCGCAACCGTGCAGCTGACCTGCGACAAGACTTCAACCTGATGGAGCAGAAG
      P  P  P  S  Q  G  Q  H  Y  F  D  R  F  S  E  D  D  P  E  Y  L  R  L  R  N  R  A  A  D  L  R  Q  D  F  N  L  M  E  Q  K
   341                                                                                                              51
     AAACGGGTCACCATGATCCTGCAGAGCCCTTCTTTCAGGGAGGAGCTGGAAGGCCTCATCCAGGAACAGATGAAGAAGGGTAACAACTCCTCCAACATCTGGGCCCTCCGACAGATCGCG
      K  R  V  T  M  I  L  Q  S  P  S  F  R  E  E  L  E  G  L  I  Q  E  Q  M  K  K  G  N  N  S  S  N  I  W  A  L  R  Q  I  A
   461                                                                                                              91
     GACTTCATGGCCAGCACCTCCCACGCAGTCTTCCCCAGCTTCCTCCATGAACTTCTCCATGATGACACCCATCAATGACCTCCACACTGCCGACTCCCTGAACCTGGCCAAGGGGGAGAGG
      D  F  M  A  S  T  S  H  A  V  F  P  A  S  S  M  N  F  S  M  M  T  P  I  N  D  L  H  T  A  D  S  L  N  L  A  K  G  E  R
   581                                                                                                             131
     CTTATGCGGTGCAAGATCAGCAGTGTCTACCGTCTTCTGGACCTCTATGGCTGGGCGCAGCTCAGCGACACCTATGTCACGCTGAGAGTGAGCAAGGAGCAGGACCACTTCCTGATCAGC
      L  M  R  C  K  I  S  S  V  Y  R  L  L  D  L  Y  G  W  A  Q  L  S  D  T  Y  V  T  L  R  V  S  K  E  Q  D  H  F  L  I  S
   701                                                                                                             171
     CCAAAGGGGTTTCCTGCAGTGAGGTCACGGCATCAGCTGATTAAGGTGAACATTCTAGGAGAGGTCGTGGAGAAGGGCAGCAGTTGCTTCCCAGTGGACACCACCGGCTTCAGTCTGCAC
      P  K  G  F  P  A  V  R  S  R  H  Q  L  I  K  V  N  I  L  G  E  V  V  E  K  G  S  S  C  F  P  V  D  T  T  G  F  S  L  H
   821                                                                                                             211
     TCAGCCATCTATGCGGCCAGGCCGGACGTGCGGTGTGCCATCCACCTGCACACGCCTGCACAGCAGCGGTGTCAGCTATGAAGTGTGGCCTCCTGCCTGTCTCCCATAATGCCCCTGCTG
      S  A  I  Y  A  A  R  P  D  V  R  C  A  I  H  L  H  T  P  A  T  A  A  V  S  A  M  K  C  G  L  L  P  V  S  H  N  A  L  L
   941                                                                                                             251
     GTGGGGGACATGGCCTACTATGACTTCAATGGGGAAATGGAGCAGGAAGCTGATCGAATCAACTTGCAGAAGTGCCTTGGACCCACCTGCAAGATTCTGGTGCTAAGAAACCATGGTATG
      V  G  D  M  A  Y  Y  D  F  N  G  E  M  E  Q  E  A  D  R  I  N  L  Q  K  C  L  G  P  T  C  K  I  L  V  L  R  N  H  G  M
  1061                                                                                                             291
     GTCGCCCTGGGTGACACCGTGGAGGAAGCTTTCTACAAGGTCTTCCACCTGCAGGCTGCGTGTGAGGTACAGGTGTCGGCTCTGTCCAGCGGCTGGGGGGACCGAGAACCTCATCCTCCTG
      V  A  L  G  D  T  V  E  E  A  F  Y  K  V  F  H  L  Q  A  A  C  E  V  Q  V  S  A  L  S  S  A  G  G  T  E  N  L  I  L  L
  1181                                                                                                             331
     GAGCAAGAGAAACACCGTCCGCATGAGGTGGGCTCTGTGCAGTGGGCCGGCAGCACCTTTGGGCCCATGCAGAAGAGCCGGCTGGGAGAGCATGAATTTGAAGCCCTCATGAGGATGCTC
      E  Q  E  K  H  R  P  H  E  V  G  S  V  Q  W  A  G  S  T  F  G  P  M  Q  K  S  R  L  G  E  H  E  F  E  A  L  M  R  M  L
  1301                                                                                                             371
     GACAACTTAGGCTACAGAACAGGCTATACGTACCGCCCACCCCTTCGTCCAAGAGAAAACCAAACACAAAAGTGAAGTGGAGATCCCAGCCACGGTCACTGCCTTTGTGTTCGAAGAGGAT
      D  N  L  G  Y  R  T  G  Y  T  Y  R  H  P  F  V  Q  E  K  T  K  H  K  S  E  V  E  I  P  A  T  V  T  A  F  V  F  E  E  D
  1421                                                                                                             411
     GGTGTCCCAGTCCCTGCTCTGCGCCAGCACGCCCAAAAGCAGCAGAAGGAAAAGACCCGCTGGCTTAACACTCCCAACACCTACCTGAGGGTGAATGTGGCCGACGAGGTGCAGAGGAAC
      G  V  P  V  P  A  L  R  Q  H  A  Q  K  Q  Q  K  E  K  T  R  W  L  N  T  P  N  T  Y  L  R  V  N  V  A  D  E  V  Q  R  N
  1541                                                                                                             451
     ATGGGCAGTCCCCGACCGAAGACCACGTGGATGAAGGCTGATGAAGTGGAAAAGTCCAGCAGCGGCATGCCCATACGGATTGAAAACCCAAACCAATTTGTGCCTCTCTACACTGACCCC
      M  G  S  P  R  P  K  T  T  W  M  K  A  D  E  V  E  K  S  S  S  G  M  P  I  R  I  E  N  P  N  Q  F  V  P  L  Y  T  D  P
  1661                                                                                                             491
     CAAGAAGTGCTGGACATGAGGAACAAGATTCGAGAACAAAACCGACAAGATATAAAGTCAGCCGGGCCTCAGTCTCAGCTCTTGGCCAGCGTCATCGCCGAGAAGAGCCGGAGTCCGGTA
      Q  E  V  L  D  M  R  N  K  I  R  E  Q  N  R  Q  D  I  K  S  A  G  P  Q  S  Q  L  L  A  S  V  I  A  E  K  S  R  S  P  V
  1781                                                                                                             531
     CAGCAGAGACTGCCCCCAACTGAAGGGGAAGCTTATCAGACTCCTGGGGCTGGGCAGGGGACCCCTGAGTCCTCAGGCCCACTCACCCCATGACCTTAGGTGCTGGGCTCTTGCTGCACA
      Q  Q  R  L  P  P  T  E  G  E  A  Y  Q  T  P  G  A  G  Q  G  T  P  E  S  S  G  P  L  T  P  *
  1901                                                                                   561
     GTGAAGGGCCACCACACAGGTGGCTGTGAGGTCCAGGGAGGACAATGTTGGCTCACTCCTGGGTGCACCAGATAGGGAGCAACAAGGAGCATCTGTTCATAGGAAGGCTTCTGGCCAGCC
  2021
     CCACACCTCCCGCCATGAGACTCCACGAGGCCCTCTCTCAGTGTCCTGCTGAATTCTGGTGGCCCAAGTGTTTCTGACCCCTGGTTTGGATGTAGCATCATCAGGATACAGGGAGGGGGT
  2141
     TGTGGCTAGGACTGCTGTGCTATGGGGTGGGAAGGGTAAAAAGTAGCCATACAAACGTAGTTAGCATTAGAAAGCATGACGTCACTGTTTCTGTGTTTGAGGGGATGGCATCTTGTCTAG
  2261
     GGCTTAGGAGCAGATGGCATTGCAATTCTGCCTTGCGTGGCATCATGGCTGTTCATGGCTTGTGGCAAAGGGTGTGACACTTTCCTAGACAATGGCTAATGGGACGTGCCAGGGCCCATC
  2381
     ACTCTCTGCAGAGCATGTTAACGTGATATGGCTGCAGATCAGGCAGCTGCAGGCTCGACCCTCACATCTGCTTCCTGACCTCCCCCAGCTCAGCCAAGTCAGTGAACTCTGAGCATGCCA
  2501
     TAGAATCAAGGTTAGAAATGACCCGTGATCCCAGGGGGATGAGAGATGAGCACAGGGCAGGACTGTCGTTCAGCAGCTCTGTCCTTCACAGCAGGAGTTCTCAGTGGAGTCCTCATCCTC
  2621
     CTTGGGGGTCTATCTTAGACTCCCTTTGACTCTGAAACCCGTTGAAATAGCCTATGTGCTCTGATGAGTGCAGCCCATAAAGTGGTGACCAGAAACCATGGTTGCTCTGGGCTCACCTCC
  2741
     CTGAGATCTCACTCTGTTCTCCTCCTCCGATTGAAAGTGATGCCTTCTGATAGGCCACAGTGAGGAAACTGACTTAGGGAAAGGACAGAAGGGTCTGGGTTCAGTCGTGTCTTGAGCAAG
  2861
     AGGGACCAGATGATTGAGAGCCAGGGATTCAACTGAGGTGCAGAGCAGAGGTTCTGTCAGTGCCCCTGCCAGGGCCTCAGGGCCTGGCCTCTTGGTCACCATGACTCCAGGATTCACTGC
  2981
     TTGGACTGAGAATGCTGAAGGCCCTGTATCTTATTGTTTTAATGCTGTGTGCTGAGGCTGCTCTTGAGGGACCTGGGGGAAGGCAGTGGGGGAGAGGGCTGGGAGGGGCAGCAGGGAAG
  3101
     AGGGCTGGCATGCTGCTTGCTCTAGCCAATGCTATGCTTGCTAGACTTTGTCTGACTAAATTTGGGCCACCAAATGTATGCACAGTCATGGCTTGGAAGTGTCCGTAGATGGGAATGCAA
  3221
     TAAAGGTATGTCTTCTGCTTGGAAAAAAAAAAAAAAAAAAA        3260
```

Fig. 10.2. Nucleotide sequence and predicted amino acid sequence of adducin 63 cDNA. The nucleotide sequence of this 3260-bp adducin 63 cDNA has an open reading frame encoding a polypeptide 561 amino acid residues. In-frame terminator codons in the 5′ untranslated region and the putative polyadenylation signal in the 3′ non-coding region are underlined. The nucelotide and the amino acid at positions 1770 and 528, respectively, corresponding to the polymorphic site (see Results and Fig. 10.6) are indicated in bold letters. Nucleotide and amino acid residues are numbered on the left and right, respectively.

Partial Sequences from Human Erythrocyte Adducin

The identity of the adducin 63 polypeptide deduced from the cDNA sequence was confirmed by comparison with partial amino acid sequences from human α and β adducin. The α and β adducin derived from human erythrocytes were purified and the isolated bands were subjected to amino acid sequencing. The resulting peptides are listed in Fig. 10.3A. Five of them share extensive homologies with adducin 63; in particular there is a stretch of 20 amino acids (sequence 14, Fig. 10.3A) perfectly matching with the sequence from amino acids 370–389. The peptides 10 from α subunit and 14 from β subunit (Fig. 10.3A) differ for one residue only. Moreover peptides 8 and 10 (Fig. 10.3A) are overlapping on the deduced sequence; therefore the total matching length is 28 residues. On a probabilistic basis such homology cannot be random, thus indicating that adducin 63 cDNA codes for a protein very similar to erythrocyte adducin, possibly one of the spleen forms of this protein. Whether adducin 63 cDNA codes for a protein similar to α or β subunit is still not clear. However, from fingerprint comparison it is evident that the α and β subunits of adducin share about 30% homology (Gardner and Bennett 1986; Salardi et al. unpublished data).

A   Sequences from whole adducin

```
1) M  X  Q  P  N  N  Q  T  H  N  P
2) M  X  E  L  E  R  S  A  W  R
3) M  P  A  P  W  A  E  E  A  A  P  Q
4) M  D  N  K  L  L  L  A  Q  X  A  K  L  R  T  Y
5) M  D  T  E  Q  Y  G  X  G  K  N
6) M  D  R  W  T  Y  L  Y  S  C  S  Q  G  K  R
7) M  A  R  E  L  R  E  Y  Q  E
```

Sequences from α subunit

```
8)  P  F  V  Q  E  K  T  K  H  K  S  E  V                    (385)
9)  P  R  P  K  T                                            (455)
10) M  L  D  N  L  G  Y  R  T  L  Y  T  Y  R  H  P  F  V  Q   (370)
11) S  G  P  Q  S  Q                                         (512)
12) T  I  P (N/Q) L  Y  Q  A  T  V  I  V
```

Sequences from β subunit

```
13) N  G  I/ L) N  E  Q  N  R  Q  D (V/ L) (C) S  A  G  L  Q  E (C) L
14) M  L  D  N  L  G  Y  R  T  G  Y  T  Y  R  H  P  F  V  Q  E   (370)
15) T  P  Q  E  N  Q  N (F) V  V
16) P  A  P  V  A  E  E  A  A  P  Q
17) Q  D  R
18) P  Q  P  D  K
```

B

```
LAK  G  ER  –L  MR  C–  KIS  S  V  YRL  LD   adducin (126)
LAK  P  ER  GK  MR  VH  KIS  N  V  NKR  LD   alfa-actin, human (77)
LAD  P  ER  GK  MR  VH  KIS  N  V  NKA  LD   alfa-actin, chick (78)
```

Fig. 10.3. Similarities of adducin 63 to other polypeptide sequences. (**A**) Amino acid sequence of human erythrocyte adducin peptides. Amino acid residues are underlined when identical to those in adducin 63 sequence. (**B**) Amino acid similarity among adducin 63 and α-actin-binding region from human and chicken (Baron et al. 1987). Identities in sequences are shown in boxes. The numbers at the right indicate amino acid position on the corresponding sequences.

Similarities of Adducin 63 to Other Protein Sequences

A database search performed with the FastDB program revealed no highly
similar sequences or obvious local similarities. It was noted, however, that the
majority of the highest ranking sequences were filamentous proteins such as
neurofilament triplet H protein, myosin heavy chain and filamentous
hemagglutinin (data not shown). A detailed search for local similarities revealed
that adducin shares one pattern in common with actin-binding proteins. As
shown in Fig. 10.3B this pattern is similar to α-actin from human and chicken
(Baron et al. 1987) and maps to the N-terminal actin-binding region of these
proteins. This region of α-actins has a sequence similarity to the actin-binding
regions of other F-actin cross-linking proteins such as porcine dystrophin and the
slime mold gelation factor (Doegel et al. 1989).

Tissue Distribution of Adducin-Like mRNA

Northern blots (Fig. 10.4) of poly(A$^+$) RNA prepared from different rat tissues
was performed to analyze the transcription products of the rat adducin gene
using SNOC 31 as probe. Liver, kidney, spleen and heart showed the same

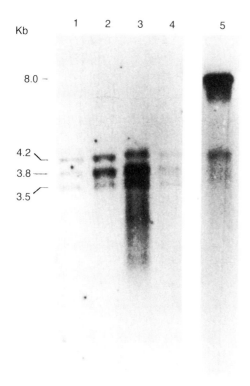

Fig. 10.4. Northern blotting analyses of adducin transcripts in different rat tissues. Ten micrograms
of poly(A$^+$) RNA isolated from kidney (lane 1), heart (lane 2), liver (lane 4), brain (lane 5) and 5 μg
from spleen (lane 3) were separated on denaturing agarose gel, transferred to nitrocellulose filter
and hybridized with ^{32}P-labeled SNOC 31 cDNA probe. The autoradiogram was exposed for ten
days. The estimated size of transcripts are indicated at the side.

pattern of hybridization with three transcripts of about 3.5, 3.8, 4.2 kb present in different isoforms in all four tissues. The smaller transcript is consistent with the size of the isolated cDNA (3242 nucleotides plus poly(A) tail). In brain a similar pattern was observed for the small RNAs, with the addition of a major brain-specific transcript of 8.0 kb. Its size was deduced by a parallel hybridization with a fibronectin probe whose mRNA is 7.9 kb (Kornblihtt et al. 1985) (data not shown).

Brain Adducin cDNA Alternative Splicing

To test whether alternative splicing is responsible for differences in adducin mRNA and protein size we used the PCR to screen different regions of rat brain adducin mRNA for lost or gained exons. Two pairs of oligonucleotide primers (SPL 3S/GAT 2, GAT 8/GAT 11), derived from the adducin 63 complementary DNA sequence were designed to amplify overlapping segments spanning the adducin 63 coding sequence (Fig. 10.5A). Amplified fragments visualized on agarose gel were Southern blotted and hybridized with specific adducin 63 cDNA probes. PCR experiments performed with rat brain cDNA and GAT 8/GAT 11 oligonucleotide primers gave three fragments: one of the size predicted from adducin 63 cDNA sequence analysis (Fig. 10.5A) and two 663 and 938 bp smaller (Fig. 10.5B and C, respectively). The whole PCR products were cloned and their nucleotide sequence determined. Only two of the 24 brain cDNA clones analyzed were different from adducin 63. It can then be inferred that these alternative forms are probably a minor species in brain. If we assume

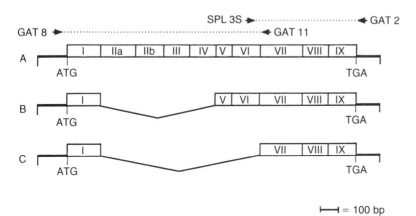

Fig. 10.5. Brain adducin alternative splicing: PCR strategy. (**A**) A schematic diagram of adducin 63 indicating exon organization as deduced by comparing the cDNA sequence with a human adducin genomic clone spanning the same region (Tisminetzky et al. in preparation). Exons are indicated by boxes and numbered in a progressive way. Non-translated regions are indicated by thin lines. (The study of exons IIa and IIb is not completed and the presence of other exons in that region cannot be excluded.) Rat brain cDNAs were amplified using two pairs of oligonucleotide primers (SPL 3S/GAT 2, Gat 8/GAT 11) to amplify overlapping segments spanning the adducin 63 coding sequence. These resulted in three distinct cDNA clones encoding the complete adducin 63 (**A**), a peptide without exons IIa, IIb, III, IV (**B**) and a peptide without exons IIa, IIb, III, IV, V, VI (**C**). In **B** and **C** the deletions are indicated by bent lines.

that the human and rat adducin genes have the same exon–intron structure in this region (Tsminetzky et al. unpublished), fragments B and C result from the loss of at least four and six exons, respectively (Fig. 10.5). The loss of exons in fragments B and C maintained the reading frame deduced from adducin 63 cDNA and no replacement or new nucleotide sequence was observed.

The alternative splicing forms described are by no means the only ones possible; large insertions would have been missed by their low yield in the amplification procedures. Other variants may have been present in lower concentrations, insufficient to be seen as amplified bands or isolated as recombinant clones.

Difference Between MHS and MNS Adducin 63

To confirm that the antigenicity of MNS erythrocyte ghosts in MHS rats was due, at least in part, to an altered adducin sequence a systematic sequencing of MNS and MHS adducin 63 mRNA was performed. The complete coding region of adducin 63 was synthesized by PCR from MNS and MHS rat spleen cDNA as two overlapping fragments. The comparison between MNS and MHS nucleotide sequence reveals a single adenine–guanine replacement in the codon for amino acid 528, resulting in a glutamine–arginine substitution. This adenine–guanine transition, found in several clones sequenced, results in a restriction fragment length polymorphism (RFLP) due to an additional HpaII site in MHS clones. In order to screen a large number of MNS and MHS rats, we designed two ogligonucleotide (GAT 1/GAT 2) to amplify enzymatically the exon IX of the adducin gene (Fig. 10.6A) and visualize the HpaII polymorphism as a restriction fragment of 53 bp for the adenine and 46 bp for the guanine substitution (Fig. 10.6B and C, respectively).

This RFLP was studied in 20 MHS and 38 MNS rats. All the MHS rats were homozygous for the guanine substitution, while MNS rats showed the three possible genotypes, suggesting a heterozygosis for adducin in this strain, which was not eliminated by the decades-long selection for high and low blood pressure.

Discussion

The complete rat adducin 63 cDNA sequence has been determined and its amino acid sequence deduced from a series of overlapping clones. This sequence has no obvious homologues in the databases. Similarities were detected to filamentous proteins such as neurofilament triplet H protein, myosin heavy chain, filamentous hemagglutinin and local similarities with actin-binding protein. A very close homology to the region involved in actin–actin interaction is present in exon IIa. The biological significance of these findings has to be corroborated by further experimentation. The relationship between adducin 63 and erythrocyte adducin has been formally established by the identity of peptide sequence from the human erythrocyte adducin with the predicted amino acid sequence of adducin 63. In fact peptides 8 and 9 (Fig. 10.3) derived from the α

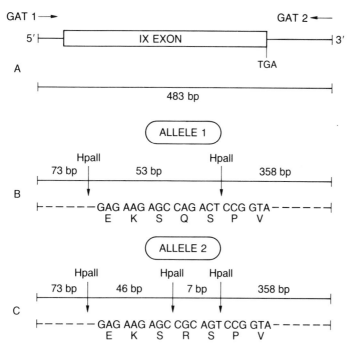

Fig. 10.6. Schematic representation of HpaII RFLP. In **A** rat genomic DNA was amplified using GAT 1/GAT 2 oligonucleotide primers corresponding to adducin exon IX flanking region sequences. After PCR, samples were HpaII digested, end labeled and submitted to electrophoresis in 12% non-denaturing acrylamide gel. Two alleles were detected: allele 1 (in **B**) with two HpaII sites corresponding to DNA fragments of 73, 53, 358 bp and allele 2 (in **C**) with an additional HpaII site in the 53-bp fragment generating a 46-bp and a 7-bp fragment. This RFLP is due to an adenine–guanine transition resulting in a glutamine (codon CAG) to arginine (codon CGG) substitution.

subunit of human adducin are identical and peptides 10 and 11 are very similar to their rat adducin 63 counterpart, while peptide 14 from the β subunit of human adducin is identical in the adducin 63 sequence (Fig. 10.3A). It should be noted that peptide 10 presents a mutation from L to G between the α and β chain. This may not be of any relevance as the adducin was prepared from a pool of several individuals and hence may be a population polymorphism. We do not know the molecular weight of the mature adducin 63 after all the post-translational modification which the primary translation product may be subject to in the cell and hence we cannot ascribe this structure to the proteins seen in erythrocyte or spleen extract. This problem is complicated by the fact that erythrocyte adducin seems to be very sensitive to protease degradation and it is difficult to differentiate between adducin degradation products (Gardner and Bennett 1986; Salardi et al. unpublished data) and eventual minor forms of adducin even in direct western blots of cellular extracts. However, we are certain of the existence of adducin 63 as a separate entity as its mRNA is clearly present in a wide variety of tissues. The cDNA clone sequenced, primer extension and RACE experiments determine very clearly the 5′- and 3′-ends of adducin 63 mRNA. Northern blots corroborate these results as the 3.5-kb species is present in most tissues analyzed. The adducin-like mRNA situation in rat brain is particularly interesting as this is the only tissue where the major species is an 8.0-kb mRNA. This is clearly longer than a calculated mRNA length of 5.0 kb required to codify the heavy 109-kDa adducin subunit in bovine brain described by Bennett et al. (1988), providing that rat brain mRNA does not include a large non-coding region. We are now studying the latter possibility by cloning size-selected brain cDNAs.

There is a complex RNA processing system in brain and possibly in the other tissues studied. In fact, assuming that the rat exon–intron structure is homologous to the one in human (Tisminetzky et al. unpublished), by PCR amplification of selected regions of the adducin 63 sequence, we have isolated at least three forms that coded for adducin isoforms including or lacking a selection of at least four to six exons (Fig. 10.5). In fact the cDNAs seen in Fig. 10.5 show that from exon I the sequence can be contiguous either with exon II, exon V or exon VII. All three different junctions occur after the third position of the codon, resulting in a protein isoform without extra amino acids but in contiguity between the exons. Direct RNAase protection analysis is needed to quantify the different species in each tissue. From the PCR data we can at least deduce that the major isoform has the exon sequence I, II, III, IV, V, VI, VII, VIII, IX, while the two minor forms have the exon sequence I, V, VI, VII, VIII, IX and I, VII, VIII, IX, respectively (Fig. 10.5). It is intereresting to note that the latter two forms lack the putative actin-binding site (Fig. 10.3B) and hence is possible that some adducin forms may not form complexes with actin.

The immunological and nucleotide sequence data are consistent with a difference between MNS and MHS adducin. In this work an amino acid polymorphism of glutamine–arginine in the codon for amino acid 528 has been established for the adducin 63. This could be consistent with the difference observed in the adducin 105 as the exon IX, where the mutation is located, is common to all four adducin-like mRNAs observed by Northern blot. The glutamine allele has been found only in the MNS strain, while all the MHS rats studied are homozygous for the arginine allele in that position. It is surprising that the heterozygosis has not been eliminated in this MNS highly inbred population. We can then infer that this heterozygosis is somehow preserved by the selection procedures. As in MNS the selection is carried out for lower blood pressure, it is possible that the maintenance of heterozygosis of the adducin locus is involved in the achievement of lower blood pressure level.

Acknowledgements

We are very grateful to our colleagues G. Bianchi, S. Tisminetzky, G. Tripodi, G. Borsani, A. Piscone and S. Salardi, who contributed extensively to the research work reviewed in this paper.

References

Baron MD, Davison MD, Jones P, Critchely DR (1987) J Biol Chem 262:17623–17629
Bennett V, Gardner K, Steiner JP (1988) J Biol Chem 263:5860–5869
Bianchi G, Ferrari P, Trizio D, Ferrandi M, Torielli L, Barber BR, Poili E (1985) Hypertension
    7:319–325

Bianchi G, Ferrari P, Barber BR (1990) In: Laragh JH, Brenner BM (eds) Hypertension: pathophysiology, diagnosis and management, Ch 58. Raven Press, New York

Doegel AA, Rapp S, Lottspeich F, Schleicher M, Stewart MJ (1989) Cell Biol 109:607–618

Ferrari P, Torielli L, Ferrandi M, Bianchi G (1986) In: Bianchi G, Carafoli E, Scarpa A (eds) Membrane pathology. Ann NY Acad Sci 488:561–566

Ferrari P, Barber BR, Torielli L, Ferrandi M, Salardi S, Bianchi G (1987) Hypertension 10 (suppl 1):32–36

Frohman MA, Dush MK, Martin GR (1988) Proc Natl Acad Sci USA 85:8998–9002

Gardner K, Bennett V (1986) J Biol Chem 261:1339–1348

Joshi R, Bennet V (1990) J Biol Chem 265:13130–13136

Kornblihtt AR, Umezawa K, Vibe-Pedersen K, Baralle FE (1985) EMBO J 4:1775–1759

Salardi S, Saccardo B, Borsani G, Modica R, Ferrandi M, Tripodi MG, Soria M, Ferrari P, Baralle FE, Sidoli A, Bianchi G (1989) Am J Hyp 2:229–237

Chapter 11

Genetic Control of Plasma Lipid, Lipoprotein and Apolipoprotein Levels: From Restriction Fragment Length Polymorphisms to Specific Mutations

S. Humphries, A. Dunning, Chun-Fang Xu and P. Talmud

Background

One of the recognized risk factors for the development of ischemic heart disease (IHD) is elevated plasma levels of cholesterol, carried in low-density lipoprotein (LDL) particles (Whayne et al. 1981; Durrington et al. 1986). The identification of the factors that determine plasma LDL cholesterol (LDL-C) levels is of major public health importance and has been the subject of intense study in recent years. Researchers have used a wide range of approaches to investigate the problem, including epidemiology, dietary and metabolic studies and more recently the techniques of cellular and molecular biology have been used. This review focuses on the identification of genetic factors that contribute to the between-individual differences in plasma LDL-C levels seen in the general population.

The major protein component of the LDL particle is the apolipoprotein B100 (apo B). This 550-kDa protein is produced by the liver and secreted as a constitutent of the very-low-density lipoprotein (VLDL) particle. During the metabolism of VLDL other apolipoproteins are removed, leaving apo B as the sole protein component of LDL. Apo B serves both to maintain the integrity of the LDL particle (Yang et al. 1986, 1989) and as ligand for the LDL receptor. Twin studies and path analysis have demonstrated a significant impact of genetic variation, with a heritability of LDL-C of 0.5–0.6 (Hamsten et al. 1986), and a similar high heritability for plasma levels of apo B (Hamsten et al. 1986). Family studies and complex segregation analysis have found evidence for a major gene determining levels of apo B, with both environmental factors and genes of small or intermediate effect making a contribution (Pairitz et al. 1988; Hasstedt et al. 1987).

Several genes have been identified that contribute to between-individual differences in plasma apo B levels. Defects in the LDL receptor, which occur in patients with the disorder familial hypercholesterolemia (FH), result in reduced removal of LDL and thus cause elevated levels of LDL (Goldstein and Brown 1983). Individuals carrying such mutations are rare in the population (roughly 1/500), but this inborn error of metabolism clearly identifies the receptor gene as being involved in the control of plasma lipid levels. Recent studies have also shown that common sequence variation in the receptor gene also contributes to between-individual differences in plasma lipid levels, with the PvuII polymorphism of the LDL receptor gene being associated with a small but significant effect (3%–10% of sample variance) on plasma LDL-C levels in healthy individuals (Pedersen and Berg 1988; Schuster et al. 1990a; Humphries et al. 1991). The mechanism of this association is not yet known.

A second gene determining plasma levels of apo B is the apo E gene (reviewed in Davignon et al. 1988). There are three common isoforms of apo E protein, caused by single base substitutions in the apo E gene. The most frequent form in the general population is apo E-3; apo E-2 is generated by a substitution of cysteine for arginine at residue 158, and apo E-4 by a substitution of arginine for cysteine at residue 112. Compared with apo E-3 and apo E-4, lipoproteins containing the apo E-2 isoform have reduced binding to the LDL receptor in vitro. In population studies apo E-2 has been shown to be associated with reduced plasma levels of cholesterol, LDL-C and apo B, whilst conversely, apo E-4 is associated with raised levels of these traits. Even though the E-2 form is generally associated with lower plasma lipid levels, this effect is masked in a small percentage of people, homozygous for the E-2 isoform, who develop Type III hyperlipidemia as a consequence of interaction between their apo E genotype and secondary factors (Utermann et al. 1979).

The Apo B Protein

The complete sequence of the apo B protein was determined following the cloning of the gene (Knott et al. 1986). The amino acids comprising the region of apo B that interacts with the LDL receptor have been defined by several approaches: firstly by homology to the apo E receptor-binding domain (Knott et al. 1986); secondly by the ability of certain monoclonal antibodies raised to apo B to block binding of LDL to the LDL receptor (Milne et al. 1989); and thirdly by the capability of lipoprotein particles containing synthetic peptides of specific apo B sequences to bind to a normal LDL receptor (Yang et al. 1986). Taken together, these methods suggest that two regions are involved in receptor binding, one spanning amino acids 3147–3157, and a second comprising residues 3345–3381 (Fig. 11.1).

This putative receptor binding region has been further expanded by the discovery and characterization of familial defective apo B100 (FDB) (Innerarity et al. 1987; Soria et al. 1989). FDB is a rare (1/000), co-dominantly inherited disorder, generated by a substitution of glutamine for arginine at residue 3500 of apo B. This mutation severely reduces binding of LDL-apo B to the LDL receptor and thus causes moderate to severe hypercholesterolemia in carriers

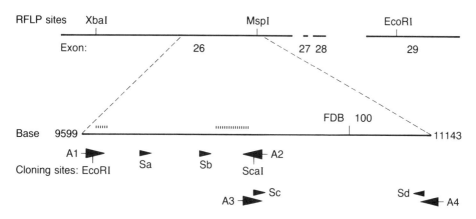

Oligo sequences, 5′ to 3′:-

Amplimers		Sequimers	
A1	TTGAAGGAATTCTTGAAAACGACAAAGCAA	SA	CAAATTCCTGGATACACTGT
A2	CACTTCCATATTTTTCGTGGTTAAGCTCAC	Sb	GAATACCAATGCTGAACTTT
A3	CATAACAGTACTGTGAGCTTAACCACGAAA	Sc	CACGAAAAATATGGAAG
A4	AAGGATCCTGCAATGTCAAGGTGTGCCTTT	Sd	TTTTCTTGGTCATTGGA

Fig. 11.1. Map of the 3′-end of the apo B gene, showing the positions and sequences of oligonucleotides used for PCR and sequencing. The bold dotted lines indicate the nucleotides encoding protein regions with homology to the apo E receptor-binding domain (nucleotides 9648–9681 and 10245–10353). The relative positions of base changes creating RFLP sites and the apo B_{3500} mutation are shown.

(Vega and Grundy 1986; Tybaerg-Hansen et al. 1990; Schuster et al. 1990b). Residue 3500 is 120 amino acids to the C-terminal side of the initially defined receptor-binding domain, but clearly affects the interaction of apo B with the receptor.

Although there is little evidence for any tertiary structure of apo B in the LDL particle, it is of interest that monoclonal antibodies to epitopes in the region of amino acids 180–670 appear to interfere partially with (though not to block) binding of LDL to the LDL receptor (Milne et al. 1989). Recently a form of apo B has been reported (Talmud et al. 1989) that is truncated at amino acid 4034 (apo B89). The patient carrying this mutation has hypobetalipoproteinemia due to a faster than normal clearance by the LDL receptor of apo B89-containing LDL particles (Krul et al. 1989). Both of these observations thus support the idea that sequence changes outside the region 3100–3500 of the apo B protein may also affect the affinity of LDL for the receptor.

The Apo B Gene

The gene for apo B has now been cloned by a number of groups, and several restriction fragment length polymorphisms (RFLPs) of the apo B gene have been identified (reviewed in Humphries 1988). Using these RFLPs, population

Table 11.1. Reported association between apo B XbaI RFLP and plasma lipid levels in healthy individuals (X− = absence of cutting site)

Origin	No.	Comments	Reference
Boston	84	X− lower chol. (NS)	Hegele et al. (1986)
London	81	X− lower chol. (lower trig.)	Law et al. (1986)
Norway	56	X− lower chol. (lower trig.)	Berg (1986)
London	62	X− lower chol. (lower trig.)	Talmud et al. (1987)
London	102	X− lower HDL	Myant et al. (1989)
Finland	176	X− lower chol. (lower LDL)	Aalto-Setala et al. (1988)
Sweden	186	X− lower chol. (NS)	Darnfors et al. (1989)
Austria	118	X− lower chol. (NS)	Paulweber et al. (1990)

NS, not significant

studies have shown that variation at a polymorphic XbaI site within the apo B gene is associated with differences in serum cholesterol levels, with the X+ allele (presence of the XbaI cutting site) being associated with raised levels. Although this association has not been seen in every study, it has been observed in many different laboratories, using samples from different countries (Table 11.1), and the consistent nature of this finding makes it highly unlikely that the observation is due to chance alone. The size of the effect associated with this RFLP is modest, explaining 3%–8% of the sample variance in cholesterol levels, and is thus of the same order of magnitude as is seen with variation at the apo E gene. The effect has not been so apparent in patients with IHD (e.g. Hegele et al. 1986; Paulweber et al. 1990), perhaps because other genetic or environmental factors have masked the impact associated with the polymorphism. Some samples from populations of different ethnic background (such as from Japan) have also not shown the association (Aburanti et al. 1988) but this may in part be because the lower frequency of the RFLP in such populations reduces the power of the sample to detect a significant effect. Similarly, studies using small samples may have failed to detect a significant association, especially if confounding variables such as gender, age and body mass index are not taken into account, due to the inclusion of individuals in the sample who are hyperlipidemic.

In some of the studies, the largest effect associated with the XbaI polymorphism has been on plasma levels of HDL cholesterol or apo A-I (Myant et al. 1989; Tikkanen et al. 1990). Although this has not been consistently observed, it again appears to have occurred more often than expected by chance alone, with the X+ allele being associated with higher levels than the X− allele. Although the mechanism of this association is puzzling, it is well known that there is an inverse relationship between the rate of VLDL metabolism and levels of HDL cholesterol (Fidge et al. 1980; Magill et al. 1982). Plasma levels of triglyceride and HDL cholesterol are inversely related, and it is possible that variation in the apo B sequence may alter the rate of lipolysis, or the rate of transfer of particle core lipids, such as cholesterol esters, between apo B-containing and apo A-I-containing particles. This exchange is mediated by the cholesterol ester transfer protein (CETP), and it has been reported recently that free fatty acids such as sodium oleate can modulate this exchange in vitro (Barter et al. 1990). If sequence variation in the apo B protein influences the levels of free fatty acids at the surface of the HDL particle, it may modulate this

CETP-mediated transfer of cholesterol esters to A-I-containing particles, and alter HDL cholesterol levels in the plasma.

These studies thus demonstrate that common variation at the apo B gene locus is involved in the determination of serum lipid levels in healthy individuals, but the molecular basis of this association is unknown. The base change that creates the XbaI site does not alter an amino acid, as it is in the third (wobble) position of a codon for threonine 2488 (Carlsson et al. 1986), and so cannot be the direct cause of the effects seen. We thus postulate a second sequence variation, elsewhere in the gene and in linkage disequilibrium with the XbaI RFLP, that leads to altered serum lipid levels. This sequence difference could occur in the promoter region of the apo B gene, where it may alter transcription of the gene and thus affect production of the protein from the intestine or liver. Alternatively, as with apo E-2 and the apo B-3500 mutation, it may change an amino acid in the protein and alter the affinity of LDL for its receptor. Evidence in favour of the latter mechanism has come from two in vivo studies that have shown that, compared with the X− allele, the X+ allele is associated with a reduced fractional catabolic rate (FCR) of autologous LDL (Demant et al. 1988; Houlston et al. 1988).

These data thus predict that LDL from individuals with the genotype X−X− would bind to the LDL receptor with increased affinity compared to LDL from individuals with the genotype X+X+. Support for this hypothesis has been obtained using a competitive binding and internalization assay of radiolabeled LDL to normal fibroblasts (Series et al. 1989). In this study "X+" LDL from older individuals (>40 years) competed poorly compared to "X−" LDL, as expected from the turnover data, but LDL from younger individuals (20 years) did not show this difference. This result is puzzling, and suggests that age-associated changes in the lipid composition of LDL particles may determine, in part, their binding to the LDL receptor. Age-related changes in lipid metabolism have been described previously, but the precise mechanism of this effect is unclear. With this reservation, the data are compatible with the hypothesis that a common amino acid change in the apo B protein may affect the affinity for the LDL receptor.

From these data we can make several predictions as to the nature of this functionally important sequence variation. Firstly, it should be a common sequence change (relative allele frequency >0.10) showing strong linkage disequilibrium with the XbaI polymorphism and allelic association of the "cholesterol-raising" allele with the X+ allele and/or the "cholesterol-lowering" allele with the X− allele. Secondly, the amino acid variation must alter structural features of the apo B protein in order to change, for example, the affinity of the particle for the LDL receptor. The location of the amino acid variation is thus likely to be in a part of the protein that is exposed on the surface of the particle. It is also likely that the residue will be physically near the sequences known to interact with the receptor (i.e. 3100–3500) although amino acid changes at a distance from this region may also affect binding through tertiary interactions. Thirdly, we would predict that LDL particles, containing apo B with the amino acid variation, would bind with different affinity to the LDL receptor, and that individuals with different genotypes will have small but consistent differences in plasma cholesterol and LDL-C and apo B levels. Because of the reported associations in some studies between the XbaI RFLP and levels of HDL-C and apo A-I, individuals with different genotypes may also show differences in levels of these traits. In any sample, all of the effects on lipid traits associated with the XbaI RFLP should be explained by the amino acid variation (i.e. the effects of the two should be of similar size and not statistically independent).

Apo B Protein Polymorphism

Several protein polymorphisms of apo B are known (Fig. 11.2). These were first identified in 1961 as antigenic determinants (Ag) using antibodies found in multiple transfused individuals (Allison and Blumberg 1961). Five pairs of epitopes were identified by these studies, with each pair representing alleles at closely linked loci (reviewed in Breguet et al. 1990). With the cloning and

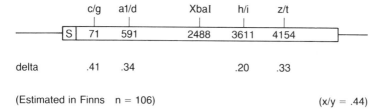

Fig. 11.2. Map of the apo B gene showing the relative position of the identified apo B Ag polymorphisms, the signal peptide polymorphism, and the XbaI RFLP variable site. The linkage disequilibrium estimate between XbaI and the protein polymorphisms in 106 healthy Finns is shown. (Data from Xu et al. 1989, 1990.)

sequencing of the apo B gene, the molecular basis of four of these polymorphisms has been determined (Table 11.2). All are the result of single base changes in the gene that cause single amino acid substitutions in the apo B protein. All of these polymorphisms can be detected at the DNA level by the presence or absence of the cutting site for a particular restriction enzyme (Table 11.2).

Table 11.2. Molecular basis of the identified apo B Ag protein polymorphisms

A (G)	Freq.	AA	No.	TR	Change	DNA	Reference
c	0.26	Ile	71	Y	Hydro.	ATC	Ma et al. (1989)
g		Thr			Polar	ACC	
a1	0.45	Val	591	Y	Hydro.	GTT	Wang et al. (1988)
d		Ala			Hydro.	GCT	
h	0.06	Gln	3611	N	Polar	CAG	Xu et al. (1989)
i		Arg			+	CGG	
z	0.16	Lys	4154	N	+	AAA	Ma et al. (1987)
t		Glu			−	GAA	Dunning et al. (1988b)

TR, peptide trypsin releasable from LDL (Yang et al. 1989)

Two of these, c/g and a1/d, are located in the N-terminal region of the protein which is trypsin accessible (Yang et al. 1989). The c/g epitopes are the result of an Ile–Val substitution at amino acid 71 (Ma et al. 1989) and the monoclonal antibodies MB19 (Tikkanen et al. 1986) and BIP45 (Duriez et al. 1987) bind strongly to the "c" epitope. The epitopes have been associated with differences in plasma cholesterol levels (both LDL-C and HDL-C) in a large study of children from Finland (Tikkanen et al. 1988). This association was not observed in an earlier smaller study (Young et al. 1987) where no adjustment for concomitants such as age, gender and BMI were taken into account. Thus, further investigations are required to confirm this effect. No association between the a1/d epitopes and lipid levels has been reported.

The two other Ag epitope pairs h/i and t/z that have been identified occur in the C-terminal part of the protein, in peptides that are not released by digestion of LDL by trypsin (Yang et al. 1989). The h/i polymorphism at residue 3611 is of interest because, like the apo B3500 mutation, it is the result of a change from Arg to Gln. This amino acid region is likely to be located on the surface of the LDL particle since it alters an epitope recognized by antibodies and is close to the region of apo B that interacts with the LDL receptor. However, unlike apo B3500, Arg–Gln 3611 does not appear to alter the affinity of the particle for the LDL receptor, and has not been consistently associated with a significant effect on lipid levels (Rajput-Williams et al. 1988; Xu et al. 1989).

The t/z polymorphism is unique in that it substitutes a positively charged amino acid Lys for a negatively charged Glu. Associations have been noted between this polymorphism and differences in VLDL triglyceride levels (Paulweber et al. 1990) and plasma lipid levels (Rajput-Williams et al. 1988) although this association has not been confirmed in other studies (e.g. Myant et al. 1989). However, many reports have noted a higher frequency of the allele distinguished by the absence of the EcoRI cutting site (t allele) in patients with IHD (e.g. Hegele et al. 1986; Myant et al. 1989; Paulweber et al. 1990). The mechanism of this association with IHD remains unclear.

At the present time the position of the remaining Ag epitope pair x/y has not been determined. This polymorphism has been associated with differences in plasma cholesterol and triglyceride levels (Berg et al. 1976), but the effect is small, and only reached statistical significance after combining data from a large number of individuals ($n = 1087$) from different studies. On average, individuals carrying one or more "x" alleles had cholesterol levels roughly 3% lower and triglyceride levels 16% lower than those with only the "y" allele. The effect on LDL-C levels was not marked, and that associated with cholesterol and triglyceride levels was strongest in samples of middle-aged individuals.

Linkage Disequilibrium

These data suggest that, of the apo B protein polymorphisms identified, the most likely candidate for the functionally important variation in linkage disequilibrium with the XbaI RFLP is the Ag (x/y) epitope. Linkage disequilibrium is a population-based concept, and is detected as association (in gametes) between two genetic markers at frequencies different from that expected from the product of their individual frequencies. Strong linkage disequilibrium has been reported in several studies between the x/y epitope pair and the XbaI RFLP (Berg et al. 1986; Xu et al. 1989). However, as shown in Fig. 11.2, there is significant linkage disequilibrium between many of the protein and DNA polymorphisms in this gene (Xu et al. 1989; 1990), and this complicates the analysis and reduces the strength of the inference that can be made.

The strength of the linkage disequilibrium detected between two polymorphisms is highly dependent on the frequency of the alleles, the size of the sample, and on the phase of the allelic association (see Thompson et al. 1988). Estimates may also be confounded by admixture in the sample of individuals of different ethnic backgrounds. Linkage disequilibrium between two markers decays because of recombination events occurring at meiosis. The rate of recombination between two markers is thus partly dependent on the distance between the two, but is also affected by the presence of sequences in the DNA which act as "hot spots" for recombination. These two factors are most apparent over large genetic distances (i.e. several million base pairs), and over shorter distances of a few thousand base pairs are primarily dependent on recent evolutionary history. Thus, if a sequence change creating a marker such as an RFLP or amino acid change occurs on an allele in the population, identified by a particular DNA or protein polymorphism, allelic association between the two may be detected.

As well as being a factor in the identification of a functionally important sequence, linkage disequilibrium may also confound the identification of such a sequence variation. An observed sequence variation may be neutral in effect, but in strong linkage disequilibrium with a second, as yet unidentified functional change elsewhere in the gene. The absolute proof of the significance of an identified change depends therefore on creation of the relevant sequence by site-directed mutagenesis, expression of the resulting protein, and an assay of the function of the protein by binding studies, or by using a monoclonal antibody.

Apo B Gene Sequencing

We have used several different techniques to compare the sequences of the apo B gene coding for amino acids 3130–3630 from eight hyperlipidemic patients (Dunning et al. 1991). Three of these patients have LDL with substantially reduced affinity for the normal LDL receptor, using an assay of binding to U937 cells (Frostegard et al. 1990), and the others have a low fractional catabolic rate of autologous LDL (Demant et al. 1988). Seven of these eight patients are homozygous for the XbaI X+ allele, and we anticipate that several of these patients will share the postulated common lipid-raising sequence difference that is in linkage disequilibrium with this allele.

The entire 1544-bp gene region studied in these patients was amplified using two overlapping sets of polymerase chain reaction (PCR) primers, designated A1/A2 and A3/A4 (Fig. 11.1). Samples were initially examined by PCR amplification (Saiki et al. 1988), using oligonucleotides A1/A2, followed by cloning into an M13 vector and sequencing. Ten clones from each subject were sequenced across this 800-bp stretch using three primers: the M13 universal primer, Sa and Sb (Fig. 11.1). Although many PCR errors were found (Dunning et al. 1988a), no base change was seen to occur in more than one clone from any patient or in more than one patient. Since a heterozygous base change would be expected to occur in approximately half the clones derived from one patient, we concluded that none of the patients studied had a mutation in this 800-bp region.

A search was made for sequence differences between bases 10 355 and 11 143 in all patients and the two positive control samples, by PCR amplification using oligonucleotides A3/A4 followed by chemical mismatch cleavage (Montandon et al. 1989). Hydroxylamine modifies mismatched deoxycytidine in the "probe" DNA, which corresponds to a nucleotide other than guanosine in the appropriate position, on the opposite strand in the test DNA while osmium tetroxide modifies mismatched deoxythymidine. A typical result is shown in Fig. 11.3, which includes as a positive control a DNA sample from an individual heterozygous for the C–T base change that creates the apo B-3500 mutation. This sample shows the expected mismatch in the hydroxylamine track, with cleavage 412 bp from the end of the PCR product. No other bands corresponding to any sequence differences were seen in any of the samples. We therefore conclude that the sequence variation in linkage disequilibrium with the XbaI RFLP is outside the region of the apo B gene coding for amino acids 3130–3030.

Maeda et al. (1988) have demonstrated similar results in swine, where an allele of apo B termed Lpb5.1 is associated with reduced clearance of LDL, hyperlipidemia and atherogenesis. These workers sequenced the region of the gene coding for amino acids 3133–3497 (the putative receptor-binding domain in swine apo B) and found no differences between this allele and three other alleles

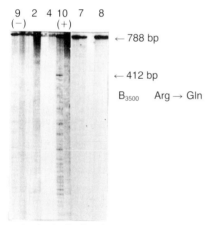

Fig. 11.3. Example of chemical mismatch cleavage analysis of samples amplified with amplimers A3/A4. Patients' code numbers are given along the top. There are two tracks per patient: the hydroxylamine reaction (H), and the osmium tetroxide reaction (O). The uncleaved, full-length PCR gives a band of 788 bp. DNA from patient 9 (genotype X− X−) was used as the "probe" for all the other samples and as the negative control (−). DNA from patient 10 (+) gives a 412-bp band arising from the base change creating the apo B3500 mutation.

which were not associated with the same phenotype. This implies that sequences outside this region must also affect interaction of LDL with the LDL receptor in pigs.

It is unlikely that the techniques used in this study were not sensitive enough to detect sequence differences between the samples examined. The PCR "errors" observed in the cloning and sequencing experiments demonstrate that single base changes can be detected using this method. Using the chemical mismatch cleavage assay, it was possible to detect the C–T base change that creates the apo B3500 mutation in one of the patients who is heterozygous for this sequence difference; this demonstrates that the mismatch technique is also sufficiently sensitive for our purpose.

Conclusions

The consistent results obtained in the XbaI RFLP association studies allow us to make a firm conclusion that variation at the apo B locus is one of the genetic factors determining plasma levels of lipids, lipoproteins and apoproteins. The size of this effect is modest, but in terms of the general population the effect is statistically and probably biologically important. The mechanism of the association is probably through an amino acid change in the apo B protein that alters LDL particle binding to the LDL receptor, or metabolism of the particle in some other way. The precise molecular determination of this association is important in the development of our understanding of apo B structure–function relationship.

Of the protein polymorphisms identified so far, the Ag (x/y) is the most likely candidate for the functionally important amino acid change. Our studies on the region of the gene coding for amino acids 3130–3630 revealed no common sequence differences, even though the study included several individuals whose LDL showed markedly reduced binding to the LDL receptor. We can thus make the strong inference that amino acid sequences outside this region must modulate LDL particle metabolism, presumably through binding to the LDL receptor.

Future work will involve the investigation of regions of the gene 3′ to those encoding the orginally envisaged receptor-binding domain since evidence suggests that the 3′-end of the apo B gene may encode the other domains of apo B involved in receptor binding that we postulate in the light of the results presented here.

Acknowledgements

Support for this work was provided by the British Heart Foundation (grant RG 5), and the Charing Cross Sunley Research Trust. The authors would like to thank Ms Elaine Osman for help in preparing the manuscript.

References

Aalto-Setala K, Tikkanen MJ, Taskinen M-R, Nieminen M, Holmberg P, Kontula K (1988) XbaI and c/g polymorphisms of the apolipoprotein B gene locus are associated with serum cholesterol and LDL-cholesterol levels in Finland. Atherosclerosis 74:47–54

Aburanti H, Matsuomoto A, Itoh H et al. (1988) A study of DNA polymorphism in the apolipoprotein B gene in a Japanese population. Atherosclerosis 72:71

Allison AC, Blumberg BS (1961) An isoprecipitation reaction distinguishing human serum protein types. Lancet i:634–637

Barter PJ, Chang LBF, Rajaram OV (1990) Sodium oleate promotes a redistribution of cholesteryl esters from high to low density lipoproteins. Atherosclerosis 84:13–24

Berg K (1986) DNA polymorphism at the apolipoprotein B locus is associated with lipoprotein level. Clin Genet 30:515–520

Berg K, Hames C, Dahlen G, Frick MH, Krishan I (1976) Genetic variation in serum low density lipoproteins and lipid levels in man. Proc Natl Acad Sci USA 73:937–940

Berg K, Powell LM, Wallis SC, Pease R, Knott TJ, Scott J (1986) Genetic linkage between the antigenic group (Ag) variation and the apolipoprotein B gene: assignment of the Ag locus. Proc Natl Acad Sci USA 83:7367–7370

Breguet G, Butler R, Butler-Brenner E, Sanchez Mazas A (1990) A worldwide population study of the Ag-system haplotypes, a genetic polymorphism of human low-density lipoprotein. Am J Hum Genet 46:502–517

Carlsson P, Darnfors C, Olofsson S-O, Bjursell G (1986) Analysis of the human apolipoprotein B gene: complete structure of the B-74 region. Gene 49:29–51

Darnfors C, Wiklund O, Nilsson J et al. (1989) Lack of correlation between the apolipoprotein B polymorphism XbaI and blood lipid levels in a Swedish population. Atherosclerosis 75:183–188

Davignon J, Gregg RE, Sing CF (1988) Apolipoprotein E polymorphism and atherosclerosis. Atherosclerosis 8:1–21

Demant T, Houlston RS, Caslake MJ et al. (1988) Catabolic rate of low density lipoprotein is influenced by variation in the apolipoprotein B gene. J Clin Invest 82:797–802

Dunning AM, Talmud P, Humphries SE (1988a) Erros in the polymerase chain reaction. Nucleic Acids Res 16:10393

Dunning AM, Tikkanen MJ, Enholm C, Butler R, Humphries SE (1988b) Relationships between DNA and protein polymorphisms of apolipoprotein B. Hum Genet 78:325–329

Dunning AM, Houlston R, Frostegard J, Revill J, Nilsson J, Hamsten A, Talmud P, Humphries S (1991) Genetic evidence that the putative receptor binding domain of apolipoprotein B (residues 3130 to 3630) is not the only region of the protein involved in interaction with the low density lipoprotein receptor. Biochim Biophys Acta 1096:231–237

Duriez P, Butler R, Tikkanen MJ et al. (1987) A monoclonal antibody (BIP 45) detects Ag (c,g) polymorphism of human apolipoprotein B. J Immunol Methods 102:205–215

Durrington PN, Hunt L, Ishola M, Kane J, Stephens WP (1986) Serum apolipoproteins AI and B and lipoproteins in middle aged men with and without previous myocardial infarction. Br Heart J 56:206–212

Fidge N, Nestel P, Ishikawa T, Reardon M, Billington T (1980) Turnover of apolipoprotein AI and AII of high density lipoprotein and relationship to other lipoproteins in normal and hyperlipidaemic individuals. Metabolism 29:643–653

Frostegard J, Hamsten A, Gidlund M, Nilsson J (1990) Low density lipoprotein-induced growth of U937 cells: a novel method to determine the receptor binding of low density lipoprotein. J Lipid Res 31: 37–44

Goldstein JL, Brown MS (1983) Familial hypercholesterolaemia. In: Stanbery JB et al. (eds) The metabolic basis of inherited disease. McGraw-Hill, New York, pp 672–712

Hamsten A, Iselius L, Dahlen G, de Faire U (1986) Genetic and cultural inheritance of serum lipids, low and high density lipoprotein cholesterol and serum apolipoproteins A-I, A-II and B. Atherosclerosis 60:199–208

Hasstedt SJ, Wu L, Williams RR (1987) Major locus inheritance of apolipoprotein B in Utah pedigrees. Genet Epidemiol 4:67–76

Hegele RA, Huang LS, Herbert PN et al. (1986) Apolipoprotein B-gene DNA polymorphisms associated with myocardial infarction. N Engl J Med 315:1509–1515

Houlston RS, Turner PR, Revill J, Lewis B, Humphries SE (1988) The fractional catabolic rate of low density lipoprotein in normal individuals is influenced by variation in the apolipoprotein B gene. Atherosclerosis 71:81–85

Humphries SE (1988) DNA polymorphisms of the apolipoprotein genes: their use in the investigation of the genetic component of hyperlipidaemia and atherosclerosis. Atherosclerosis 72:89–108

Humphries S, Coviello DA, Masturzo P, Balestreri R, Orecchini G, Bertolini S (1991) Variation in the LDL-receptor gene is associated with differences in plasma LDL-cholesterol levels in young and old normal individuals from Italy. Atherosclerosis (in press)

Innerarity TL, Weisgraber KH, Arnold KS et al. (1987) Familial defective apolipoprotein B-100: low density lipoproteins with abnormal receptor binding. Proc Natl Acad Sci USA 84:6919–6923

Knott TJ, Pease RJ, Powell LM et al. (1986) Complete protein sequence and identification of structural domains of human apolipoprotein B. Nature 323:734–738

Krul ES, Kinoshita M, Talmud P et al. (1989) Two distinct truncated apolipoprotein B species in a kindred with hypobetalipoproteinemia. Atherosclerosis 9:856–868

Law A, Wallis SC, Powell LM et al. (1986) Common DNA polymorphism within coding sequence of apolipoprotein B gene associated with altered lipid levels. Lancet i:1301–1303

Ma Y, Schumaker V, Butler R, Sparkes RS (1987) Two DNA restriction fragment length polymorphisms (RFLPs) associated with Ag (t/z) and Ag (c/g) antigenic sites of human apolipoprotein B. Arteriosclerosis 7:301–305

Ma Y, Wang X, Butler R, Schumaker VN (1989) Bsp 12861 restriction fragment length polymorphism detects Ag (c/g) locus of human apolipoprotein B in all 17 persons studied. Atherosclerosis 9:242–246

Maeda N, Ebert DL, Doers TM et al. (1988) Molecular genetics of the apolipoprotein B gene in pigs in relation to atherosclerosis. Gene 70:213–229

Magill P, Rao SN, Miller NE et al. (1982) Relationships between the metabolism of high-density and very-low-density lipoproteins in man: studies of apolipoprotein kinetics and adipose tissue lipoprotein lipase activity. Eur J Clin Invest 12:113–120

Milne R, Theolis R Jr, Maurice R et al. (1989) The use of monoclonal antibodies to localize the low density lipoprotein receptor-binding domain of apolipoprotein B. J Biol Chem 264:19754–19760

Montandon AJ, Green PM, Giannelli F, Bentley DR (1989) Direct detection of point mutations by mismatch analysis: application to haemophilia B. Nucleic Acids Res 17:3347–3358

Myant NB, Gallagher J, Barbir M, Thompson GR, Wile D, Humphries SE (1989) Restriction fragment length polymorphisms in the apoB gene in relation to coronary artery disease. Atherosclerosis 77:193–201

Pairitz G, Davignon J, Maillouz H, Sing CF (1988) Sources of interindividual variation in the quantitiative levels of apolipoprotein B in pedigree ascertained through a lipid clinic. Am J Hum Genet 43:311–321

Paulweber P, Friedl W, Krempler S, Humphries S, Sandhofer F (1990) Association of DNA polymorphism at the apolipoprotein B gene locus with coronary heart disease and serum very low density lipoprotein levels. Atherosclerosis 10:17–24

Pedersen JC, Berg K (1988) Normal DNA polymorphism at the low density lipoprotein receptor (LDL-R) locus associated with serum cholesterol levels. Clin Genet 34:306–312

Rajput-Williams J, Knott TJ, Wallis SC et al. (1988) Variation of apolipoprotein B-gene is associated with obesity, high blood cholesterol levels, and increased risk of coronary heart disease. Lancet ii:1442–1446

Saiki RK, Gelfand DH, Stoffel S et al. (1988) Primer-directed enzymatic amplification of DNA with a thermostable DNA polymerase. Science 239:487–491

Schuster H, Rauh G, Korman B et al. (1990a) Familial defective apolipoprotein B-100: comparison with familial hypercholesterolemia in 18 cases detected in Munich. Atherosclerosis 10: 577–581

Schuster H, Humphries S, Rauh G et al. (1990b) Association of DNA-haplotypes in the human LDL-receptor gene with normal serum cholesterol levels. Clin Genet 38:401

Series J, Cameron I, Caslake M, Gaffney D, Packard CJ, Shepherd J (1989) The XbaI polymorphism of the apolipoprotein B gene influences the degradation of low density lipoprotein in vivo. Biochim Biophys Acta 1003:183–188

Soria LF, Ludwig EH, Clarke HR, Vega GL, Grundy SM, McCarthy BJ (1989) Association between a specific apolipoprotein B mutation and familial defective apolipoprotein B-100. Proc Natl Acad Sci USA 86:587–591

Talmud PJ, Barni N, Kessling AM et al. (1987) Apolipoprotein B gene variants are involved in the determination of serum cholesterol levels: a study in normo- and hyperlipidaemic individuals. Atherosclerosis 67:81–89

Talmud P, King Underwood L, Krul E, Schonfeld G, Humphries S (1989) The molecular basis of truncated forms of apolipoprotein B in a kindred with compound heterozygous hypobetalipoproteinemia. J Lipid Res 30:1773–1779

Thompson EA, Deeb S, Walker D, Motulsky AG (1988) The detection of linkage disequilibrium between closely linked markers. Am J Hum Genet 42:113–124

Tikkanen MJ, Ehnholm C, Butler R, Young SG, Curtiss LK, Witztum JL (1986) Monoclonal antibody detects Ag polymorphism of apolipoprotein B. FEBS Lett 202:54–58

Tikkanen MJ, Viikari J, Kakerblom HK, Pesonen E (1988) Apolipoprotein B polymorphism and altered apolipoprotein B and low density lipoprotein cholesterol concentrations in Finnish children. Br Med J 296:169–170

Tikkanen MJ, Xu C-F, Hamalainen T, Talmud P, Sarna S, Huttunen JK (1990) XbaI polymorphism of the apolipoprotein B gene influences plasma lipid response to diet intervention. Clin Genet 37:327–334

Tybaerg-Hansen A, Gallagher J, Vincent J et al. (1990) Familial defective apolipoprotein B-100: detection in the United Kingdom and Scandinavia, and clinical characteristics of ten cases. Atherosclerosis 80:235–242

Utermann G, Vogelberg KH, Steinmetz A, Schoenborn W, Pruin N, Joeschke M, Hees M, Canzler H (1979) Polymorphism of apolipoprotein E. II. Genetics of hyperlipoproteinaemia, Type III. Clin Genet 15:37–62

Vega GL, Grundy SM (1986) In vivo evidence for reduced binding of low density lipoproteins to receptors as a cause of primary moderate hypercholesterolaemia. J Clin Invest 78:1410–1414

Wang X, Schlapfer P, Ma Y, Butler R, Elovson J, Schumaker VN (1988) Apolipoprotein B: the Ag (1/d) immunogenetic polymorphism coincides with a T-to-C substitution at nucleotide 1981, creating an alu 1 restriction site. Atherosclerosis 8:429–435

Whayne TF, Alaupovic P, Curry MD, Lee ET, Anderson PS, Schechter E (1981) Plasma apolipoprotein B and VLDL-, LDL- and HDL- cholesterol as risk factors in the development of coronary artery disease in male patients examined by angiography. Atherosclerosis 39:411–424

Xu C-F, Nanjee N, Tikkanen MJ et al. (1989) Apolipoprotein B amino acid 3611 substitution from arginine to glutamine creates the Ag (h/i) epitope: the polymorphism is not associated with differences in serum cholesterol and apolipoprotein B levels. Hum Genet 82: 322–326

Xu C-F, Tikkanen MJ, Huttunen JK et al. (1990) Apolipoprotein B signal peptide insertion/deletion polymorphism is associated with Ag epitopes and involved in the determination of serum triglyceride levels. J Lipid Res 31:1255–1261

Yang C-Y, Chen S-H, Gianturco SH et al. (1986) Sequence, structure, receptor-binding domains and internal repeats of human apolipoprotein B-100. Nature 323:738–742

Yang C-Y, Zi-Wei G, Weng S, Kim T-W, Chen S-H, Pownall HJ, Sharp PM, Liu S-W, Li W-H, Gotto AM, Chan L (1989) Structure of apolipoprotein B-100 of human low density lipoproteins. Atherosclerosis 9:96–108

Young SG, Bertics SJ, Scott TM et al. (1987) Apolipoprotein B allotypes MB19$_1$ and MB2 in subjects with coronary artery disease and hypercholesterolemia. Arteriosclerosis 7:61–65

Chapter 12

Receptor Regulation of Lipoprotein Metabolism

U. Beisiegel

Introduction

Lipoproteins are the carrier of all plasma lipids and they have a most important physiological function in the regulation of cholesterol and triglyceride metabolism in the body. In addition fatty acid metabolism is partly dependent on the structure of lipoproteins. For these reasons it is obvious that the regulation of lipoprotein metabolism plays a major role in energy generation in the body. Next to these physiological functions defects in lipoprotein metabolism can have severe pathophysiological consequences, like hyperlipoproteinemia and arteriosclerosis.

The regulation of lipoprotein metabolism is determined by lipoprotein synthesis as well as catabolism, which is mediated by the interaction of apolipoproteins with special lipoprotein receptors. In this chapter the role of lipoprotein receptors in lipid metabolism will be discussed, in particular directed to the question of regulation of receptor-dependent lipoprotein catabolism. It should be mentioned here that defects in lipoprotein receptors can be the cause of premature coronary heart disease, best exemplified by defects in the receptor for low-density lipoproteins (LDL) which lead to the most severe familial hypercholesterolemia in its homozygous form.

The most important lipoproteins for the transport of exogenous lipids are the chylomicrons. LDL are the main carrier of endogenous cholesterol and high-density lipoproteins (HDL) play an important role in the reverse cholesterol transport. Under pathophysiological aspects we should include oxidized LDL (oxLDL) as a main substrate for plaque development in arteriosclerosis. For all these lipoproteins specific receptors have been described.

Table 12.1. Lipoprotein receptors

Lipoprotein receptor	Structure and function	Regulation	References
LDL receptor	Fully elucidated; 115-kDa	Mainly by cellular cholesterol levels	Brown and Goldstein (1986)
Modified LDL receptor	Several concepts; structure described, candidate 220 kDa	Not known	Arai et al. (1989) Sparrow et al. (1989) Kodama et al. (1990)
HDL receptor	Several models; 110-kDa candidate	By cholesterol accumulation in cells	Graham and Oram (1987) Schmitz et al. (1985)
CR receptor	Described for a candidate (600-kDa)	Under investigation	Herz et al. (1988) Beisiegel et al. (1989)

The extent of our current knowledge on these receptors varies significantly (see Table 12.1). While the LDL receptor is fully characterized (Brown and Goldstein 1986), different concepts exist for the receptor which supposedly recognizes oxLDL (Arai et al. 1989; Sparrow et al. 1989; Kodama et al. 1990). The structure of a potential oxLDL receptor has been described recently (Kodama et al. 1990) but as yet there are no data on its regulation in the literature. Several groups have reported on potential HDL-receptor proteins which were, however, different in size (Oram et al. 1983; Graham and Oram 1987; Schmitz et al. 1985). None of those candidates has yet been proven to be the functional receptor protein, responsible for the "reverse cholesterol transfer". Therefore the final identity of the HDL-receptor has not yet been fully elucidated. As regards the regulation of the HDL receptor it has been reported that the accumulation of cholesterol in the cells increases the receptor activity.

Here I will concentrate on the receptor for chylomicron remnants (CR) – the metabolic products from the intestinal-derived chylomicrons. In the last two years there has been a rapid development in understanding the potential CR receptor (Herz et al. 1988; Beisiegel et al. 1989; Kowal et al. 1989; Lund et al. 1989). Recent data on the candidate receptor will be presented and discussed. For this discussion the LDL receptor, whose structure–function relationship and regulation are well understood (Brown and Goldstein 1986) will serve as a model system.

Separate Pathways for Exogenous and Endogenous Lipids

CR formation and catabolism can be considered as the "exogenous pathway" of lipid metabolism since dietary lipids are transported with these lipoproteins to the liver and perhaps also to other tissues (Hussain et al. 1989). This metabolic path is paralleled by the "endogenous pathway", where LDL are formed from very-low-density lipoproteins (VLDL) and catabolized via the LDL receptor. Both these pathways (Fig. 12.1) have several similarities which need to be pointed out for the discussion on the regulation of lipoprotein metabolism.

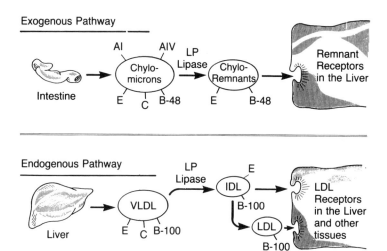

Fig. 12.1. Schematic demonstration of the pathways of exogenous and endogenous lipids. Lipolysis of the triglyceride-rich particles is performed in parallel by the endothelial-bound enzyme lipoprotein lipase. Catabolism in the liver cells is, however, mediated by different receptors.

Chylomicrons are, like VLDL, lipolyzed in the bloodstream by the endothelial-bound enzyme lipoprotein lipase (LPL). LPL activity induces several changes resulting overall in smaller particles, which are poorer in triglycerides and richer in cholesterol. These particles have been shown to be the best ligand for the receptor-mediated catabolism of chylomicrons (Cooper 1977; Floren et al. 1981). These small particles have been designated CR or, in case of the VLDL, intermediate-density lipoproteins (IDL). Lipolysis leads also to a change in apolipoprotein composition. The most extensively described and probably also most important change is the relative increase in apolipoprotein E (apo E), together with a decrease in apo C (Windler et al. 1980).

CR are then rapidly taken up by the liver while, in man, most of the IDL is further metabolized to LDL. The catabolism of LDL in the liver is mediated by the LDL receptor and this process has been fully elucidated and will be reviewed here, since it is necessary for understanding the concepts concerning characterization of the potential CR receptor.

The LDL receptor is able to bind LDL, internalize with the particle and enter the endocytotic pathway. In the so-called "compartment of uncoupling of receptor and ligand" (CURL) (Geuze et al. 1983) the receptor goes into its recycling pathway, while the LDL are degraded in lysomes. The binding properties of the receptor have been studied and it is known that it binds the large form of apolipoprotein B (apo B-100). In addition the LDL receptor has a high-affinity binding capacity for apo E. Patients with a total absence of LDL receptor activity (no expression of receptor protein) are an excellent model for the function of the LDL receptor (Brown and Goldstein 1976). These patients express a severe hypercholesterolemia due to selective accumulation of LDL in the plasma, called familial hypercholesterolemia (FH). No major abnormalities

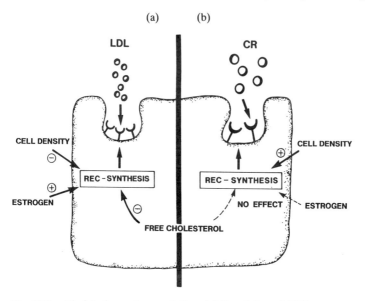

Fig. 12.2. Model of receptor regulation. (**a**) Regulation of LDL receptor synthesis: inhibition by free intracellular cholesterol, stimulation by estrogen and down-regulation by high cell density. (**b**) For LRP, the potential chylomicron remnant receptor, we found no effect by cholesterol, no effect by estrogen and a stimulating effect of high cell density.

could be seen in the chylomicron catabolism of these patients. In tissue culture studies it has been shown that the LDL receptor is down-regulated by intracellular cholesterol, stimulated by estrogen treatment, and that with high cell density receptor activity is decreased (Brown and Goldstein 1986; Fig. 8.2a). The molecular mechanism of the regulation of receptor expression by cholesterol has been recently elucidated. Cholesterol acts, via a sterol-binding protein, directly on the sterol-regulated element in the promoter of the LDL receptor gene. Receptor synthesis is repressed by this element (Smith et al. 1990).

Postulating a Chylomicron Remnant Receptor

Several studies have demonstrated that CR catabolism is mediated by apo E (Windler et al. 1980; Sherrill et al. 1980) and as mentioned above it has been shown that apo E binds with high affinity to the LDL receptor. These facts led to the concept that the LDL receptor might also be responsible for CR clearance in the liver. However, in addition to many tissue culture and animal studies there is one most striking piece of evidence that the LDL receptor is not a CR receptor in vivo: i.e. the normal CR catabolism in FH patients who have no LDL receptor activity.

Studies on Chylomicron Remnant Catabolism

There have been many studies on chylomicron catabolism over the last 30 years; however, only a few laboratories have tried to characterize the potential CR receptor. Most studies concentrated on the ligand, namely CR, and the regulation of its catabolism (Cooper 1977; Nilsson 1977; Floren et al. 1981). In summary the following is known concerning CR catabolism: chylomicrons need to be lipolyzed to act as a ligand for the potential CR receptor, and apo E is necessary for the uptake of CR in the liver; apo Cs seem to inhibit this uptake. This apolipoprotein dependence of catabolism explains the fact that in experiments concerning detection of the CR receptor we looked for apo E-binding proteins. Intracellular cholesterol levels in tissue culture experiments or cholesterol feeding in animal studies had no effect on CR catabolism. Estrogen treatment also had no influence on CR uptake.

In 1986 the first attempts were made to characterize a CR receptor protein (Hui et al. 1986). However, in this approach, as well as in later studies in our laboratory, the apo E-binding proteins which had been isolated with apo E affinity chromatography were found to have no receptor features (Beisiegel et al. 1988). Only recently could we describe a possible candidate for the long-searched-for receptor protein (Beisiegel et al. 1989). Herz et al. (1988) had described an "LDL receptor-related protein" (LRP) of 514 kDa with a high sequence homology to the LDL receptor. LRP was shown to be a protein with

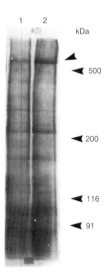

Fig. 12.3. Autoradiography of a nitrocellulose after SDS PAGE and electroblotting. HepG2 cells were metabolically labeled with [^{35}S]methionine and apo E-liposomes were linked to the "apo E-receptor" by the chemical cross-linker SADP (Pierce). The washed cells were solubilized in detergent (NP40) and an immunoprecipitation performed. In lane 1 immunoprecipitation was performed with anti-apo E (the ligand) and in lane 2 anti-LRP was used. Next to background bands which are seen in different amounts in different experiments a clear band is labeled at 600 kDa. The anti-apo E only precipitated the apo E-bound LRP at the cell surface and therefore less than the anti-LRP which precipitates all synthesized LRP.

all the structural features necessary for a lipoprotein receptor. At the same time we performed cross-linking studies. Apo E liposomes were bound to a iodinated chemical cross-linking molecule (Denny Jaffe reagent, NEN, NEX 227) and allowed to bind to its potential receptor on the surface of human hepatoma cells (HepG2). After washing, photoactivation and a cleavage step the label of the cross-linking molecule was attached to the potential receptor protein, which could now be demonstrated on an autoradiography after SDS PAGE (Fig. 12.3). We detected a 600-kDa protein as the apo E-binding protein and thereby the potential CR receptor. The immunological identity of this protein as LRP was shown in a collaborative study (Beisiegel et al. 1989). Since then LRP and its potential role in chylomicron metabolism has been under investigation in several laboratories. It is considered today to be the best candidate for the CR receptor.

Characterization of LRP as the Potential Chylomicron Remnant

Using cross-linking we have shown that LRP binds all three apo E isoforms (apo E-2, E-3 and E-4). Moreover, apo E binding in the cross-linking experiments could be suppressed by human IDL and CR. Direct cross-linking with CR is very difficult to perform due to the size of the particle, but changes in the experimental set-up which may allow such direct evidence are currently being tested. In collaboration with J. Herz we have demonstrated LRP in human hepatoma cells, mouse macrophages, human fibroblasts and human lymphocytes, as well as in human bone marrow. LRP is present in all species tested so far and it seems that LRP is a phylogenetically very old protein.

To confirm its role as a lipoprotein receptor we analyzed in collaboration with S. Jäckle the presence of LRP in rat liver endosomes. Using cross-linking agents we could demonstrate that in isolated endosomes apo E liposomes bind to LRP. This experiment confirmed data from Lund et al. (1989), who also showed the presence of LRP in the vesicles of the endcytotic pathway. Sequence data of the cytoplasmatic tail region from Herz et al. (1988) also provided evidence that LRP, like the LDL receptor, can undergo rapid endocytosis. The LRP contains a duplication of the sequence which has previously been shown to be essential for the clustering in coated pits.

In addition to the apo E binding demonstrated by cross-linking experiments in our laboratory, two other groups have tried to demonstrate the function of LRP as an apo E-binding protein. Lund et al. (1989) used the ligand blotting system to demonstrate binding of apo E-rich lipoproteins to LRP, and Kowal et al. (1989) have shown LRP-dependent uptake of apo E-enriched rabbit β-VLDL into LDL receptor-deficient human fibroblasts.

Structural Features of LRP

Structural studies by Herz et al. (1990) led to the concept that LRP undergoes proteolytic cleavage in the late Golgi. By this process two subunits (LRP-515 and LRP-85) are formed which are, however, kept together by an as yet unknown mechanism on their way to the plasma membrane. That paper elegantly clarified observations from our laboratory and others which described a "90-kDa fragment" of LRP which was recognized by the antibody against a peptide in the intracellular tail. Other multimeric cell surface proteins are known to be synthesized as single polypeptide chains and then cleaved by proteolysis to form subunits, exemplified by the receptors for insulin and IGF-I. In these cases the subunits are, however, covalently attached by disulfide bonds forming heterodimers (Olson et al. 1988).

Regulation of LRP

Due to its structural homology to the LDL receptor as well as the possible functional similarities in the endocytotic pathway, it was questioned whether the regulation of LRP is comparable to the LDL receptor. It could, however, be shown in our laboratory as well as in others (Kowal et al. 1989; Lund et al. 1989) that LRP is neither regulated by intracellular cholesterol in tissue culture nor by cholesterol feeding in rats. Estrogen was also shown not to influence LRP activity, and in our hands cell density has no inhibitory effect on the expression of LRP on HepG2 cells. On the contrary, higher cell density seemed rather to increase the binding activity (Fig. 12.2b). This indicates that the factors most important in the regulation of the LDL receptor have no effect on LRP, but regulation of LRP corresponds quite well to the regulation described for CR catabolism (see above).

A recent and totally different approach might, however, contribute to the understanding of the regulation of CR catabolism. Bihain et al. (1989) have shown that the addition of free fatty acid (particularly oleate) to the medium decreases LDL receptor activity in normal human fibroblasts. However, using the same ligand, ^{125}I-labeled LDL, on LDL receptor-deficient fibroblasts increased uptake of LDL and apo E-containing VLDL could be demonstrated in the presence of 1 mM oleate (Yen et al. 1989). Therefore a "lipolysis-stimulated receptor" (LSR) was postulated, distinct from the LDL receptor which is able to bind LDL and apo E-containing lipoproteins. The structure of the LSR has not yet been studied, but it seems to be an obvious question whether the LSR might be identical to LRP. In binding studies on HepG2 cells we were able to show that the addition of lipoprotein lipase containing post-heparin plasma stimulated the binding of chylomicrons 4–5-fold (Fig. 12.4). This indicates that the lipolysis is either changing the particles so that they are a better ligand for the LRP, or is producing stimulating factors such as free fatty acids, as described above. In further experiments we need to elucidate the mechanisms by which the binding is stimulated.

Conclusion

All features described so far for LRP are in accordance with its possible function in CR catabolism, and therefore LRP remains for the time being the best candidate for the postulated CR receptor. The earlier described regulation for

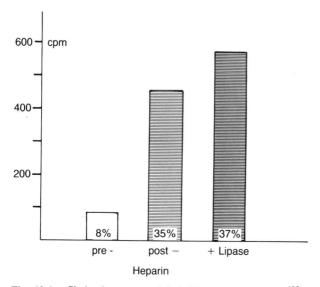

Fig. 12.4. Chylomicrons were labeled by incubation with ^{125}I-apo E. These chylomicrons were incubated in either pre-heparin plasma, post-heparin plasma (with lipase) or with purified lipase (from Dr Olivecrona, Umeå). After lipolysis or control incubation with pre-heparin plasma chylomicrons or remnants were given on HepG2 cells for binding and uptake at 37°C for 2 hours. Radioactivity associated with the cells after washing is shown in the figure. A four to fivefold increase can be seen after lipolysis. The percentages given indicate the amounts of specific binding in the presence of an excess of unlabeled chylomicron remnants.

CR clearance in the liver could also be confirmed in the regulation of this receptor protein. LRP has been shown to be neither down-regulated by intracellular cholesterol nor stimulated by estrogen, and high cell density in tissue culture did not decrease the receptor activity.

There have been many implications that lipolysis or its products might be responsible for an up-regulation of CR catabolism. In this context the concept of the LSR might play a role in CR catabolism. It seems, however, at the moment not to be meaningful to postulate LSR as another independent lipoprotein receptor.

After all the evidence demonstrated for LRP as a possible CR receptor we do not yet have the final proof for this function and future experiments will have to elucidate whether LRP is responsible for CR catabolism in vivo.

References

Arai H, Kita T, Yokode M, Narumiya S, Kawai C (1989) Multiple receptors for modified LDL in mouse peritoneal macrophages: different uptake mechanisms for acetylated and oxidized LDL. Biochem Biophys Res Commun 159:1375–1382

Beisiegel U, Weber W, Havinga JR, Ihrke G, Hui DY, Wernette-Hammond ME, Turck CW, Innerarity TL, Mahley RW (1988) Apolipoprotein E-binding proteins isolated from dog and human liver. Arteriosclerosis 8:288–297

Beisiegel U, Weber W, Ihrke G, Herz J, Stanley KK (1989) The LDL-receptor-related protein, LRP is an apolipoprotein E-binding protein. Nature 341:162–164

Bihain BE, Deckelbaum RJ, Yet FT, Gleeson AM, Carpentier YA, Witte JD (1989) Unesterified fatty acids inhibit the binding of low density lipoproteins to human fibroblast low density lipoprotein receptor. J Biol Chem 264:17316–17312

Brown MS, Goldstein JL (1976) Familial hypercholesterolemia: a genetic defect in the low density lipoprotein receptor. N Engl J Med 294:1386–1390

Brown MS, Goldstein JL (1986) A receptor-mediated pathway for cholesterol homeostasis. Science 232:34–47

Cooper AD (1977) The metabolism of chylomicron remnants by isolated perfused rat liver. Biochem Biophys Acta 488:464–474

Floren C-H, Albers JJ, Kudchodkar BJ, Bierman EL (1981) Receptor-dependent uptake of human chylomicron remnants by cultured skin fibroblasts. J Biol Chem 256:425–433

Geuze HJ, Slot JW, Strous GJAM, Lodish HF, Schwartz AL (1983) Intracellular site of asialoglycoprotein receptor–ligand uncoupling: double label immunoelectron microscopy during receptor-mediated endocytosis. Cell 32:277–287

Graham DL, Oram JF (1987) Identification and characterization of a high density lipoprotein-binding protein in cell membranes by ligand blotting. J Biol Chem 262:7439–7442

Herz J, Hammann U, Rogne S, Myklebost O, Gausepohl H, Stanley KS (1988) Surface location and high affinity. EMBO J 7:4119–4127

Herz J, Kowal RC, Goldstein JL, Brown MS (1990) Proteolytic processing of the 600 kD LRP occurs in a trans-Golgi compartment. EMBO J 9:1769–1776

Hui DY, Brecht WJ, Hall EA, Friedman G, Innerarity TL, Mahley RW (1986) Isolation and characterization of the apolipoprotein E receptor from canine and human liver. J Biol Chem 261:4256–4267

Hussain MM, Mahley RW, Boyles JK, Fainaru M, Brecht WJ, Lindquist PA (1989) Chylomicron–chylomicron remnant clearance by liver and bone marrow in rabbits. J Biol Chem 254:9571–9582

Kodama T, Freeman M, Rohrer L, Zabrecky J, Matsudaira P, Krieger M (1990) Type I macrophage scavenger receptor contains α-helical and collagen-like coiled coils. Nature 343:531–535

Kowal RC, Herz J, Goldstein JL, Esser V, Brown MS (1989) LDL receptor-related protein mediates uptake of cholesteryl esters derived from apolipoprotein E-enriched lipoproteins. Proc Natl Acad Sci USA 86:5810–5814

Lund H, Takahashi K, Hamilton RL, Havel RJ (1989) Lipoprotein binding and endosomal itinerary of the LDL receptor-related protein in rat liver. Proc Natl Acad Sci USA 86:9318–9322

Nilsson A (1977) Effects of anti-microtubular agents and cycloheximide on the metabolism of chylomicron cholesteryl esters by hepatocyte suspension. Biochem J 162:367–377

Olson TS, Bamberger MJ, Lane MD (1988) J Biol Chem 263:7342–7351

Oram JF, Brinton EA, Bierman EL (1983) Regulation of HDL receptor activity in cultured human skin fibroblasts and human arterial smooth muscle cells. J Clin Invest 72:1611–1621

Schmitz G, Robenek H, Lohmann U, Assmann G (1985) Interaction of HDL with cholesteryl ester-laden macrophages: biochemical and morphological characterization of cell surface receptor binding endocytosis and resecretion of HDL by macrophages. EMBO J 4:613–622

Sherrill BC, Innerarity TL, Mahley RW (1980) Rapid hepatic clearance of the canine lipoproteins containing only the E apoprotein by a high affinity receptor. J Biol Chem 255:1804–1807

Smith JR, Osborne TF, Goldstein JL, Brown MS (1990) Identification of nucleotides responsible for enhancer activity of sterol regulatory element in LDL receptor gene. J Biol Chem 265:2306–2310

Sparrow CP, Parthasarathy S, Steinberg D (1989) A macrophage receptor that recognizes oxidized LDL but not acetylated LDL. J Biol Chem 264:2599–2604

Windler E, Chao Y, Havel RJ (1980) Regulation of hepatic uptake of triglyceride-rich lipoproteins in the rat. J Biol Chem 256:8303–8307

Yen FT, Bihain BE, Gleeson AM, Vogel T, Gorecki M, Deckelbaum RJ (1989) Free fatty acids increase the binding of apo E-containing particles to the cultured human fibroblast. Circulation 80 (Suppl) II-487

Chapter 13

The High-Density Lipoprotein Receptor
J. F. Oram

Introduction

An early characteristic of developing atherosclerotic lesions is the accumulation of cells in the artery wall that contain numerous lipid droplets comprised mostly of cholesteryl esters. The major cells that accumulate cholesterol in these lesions are macrophages which are derived from circulating monocytes that penetrate the endothelial barrier in response to inflammatory signals. Although the precise mechanism for subendothelial recruitment of monocytes and their subsequent accumulation of cholesterol is still speculative, it is generally believed that circulating cholesterol-rich lipoproteins, namely low-density lipoprotein (LDL), infiltrate the artery wall and become oxidized in the subendothelial space (Steinberg et al. 1989). These oxidized particles may be chemotactic and promote monocyte recruitment and their differentiation into macrophages. The cell membranes of these macrophages contain unique "scavenger" receptor proteins that bind oxidized LDL and allow the cells to engulf the bound particles and degrade their lipid and protein components (Brown and Goldstein 1983). The excess cholesterol liberated by this process is esterified by intracellular enzymes to form cholesterol ester lipid droplets.

This model of atherogenesis is consistent with data from numerous studies showing a positive correlation between plasma levels of LDL cholesterol and risk for coronary heart disease, since infiltration of the artery wall with LDL particles may be a function of their plasma concentration. Although about two-thirds of the cholesterol in the blood is carried in LDL particles, most of the remaining cholesterol is carried in smaller particles called high-density lipoproteins (HDL). In contrast to LDL, population studies have shown an inverse correlation between plasma HDL levels and risk for atherosclerosis, suggesting that HDL may protect against atherogenesis. The protection may be related to the role HDL plays in "reverse cholesterol transport", a pathway by which cholesterol is transported from peripheral cells to the liver for excretion from the body (Glomset 1968). When operating efficiently, this pathway may continually clear excess cholesterol from cells of the artery wall and retard lesion formation.

An important step in the reverse cholesterol transport pathway is the actual removal of cholesterol from cells by HDL particles. Work from this author's laboratory has shown that this step involves a complex cellular pathway that is modulated by the interaction of HDL particles with specific receptor proteins on the cell surface. The experimental evidence for an HDL receptor pathway is briefly described in the following sections.

High-Density Lipoprotein Receptor Hypothesis

With the exception of steroidogenic cells, a typical extrahepatic cell utilizes cholesterol for only one purpose: as a structural component of membranes, particularly the plasma membrane that surrounds the cells. The relative cholesterol composition of membranes must be maintained within narrow limits for cells to grow and function normally. Thus cells have developed several tightly regulated pathways to ensure that the supply of cholesterol is sufficient and that membranes do not become overloaded with cholesterol when supply exceeds demand. When cholesterol is released into cellular compartments subsequent to uptake and degradation of sterol-rich lipoproteins, the excess cholesterol is either re-esterified and stored as lipid droplets, or excreted from cells to HDL particles. The importance of the excretory pathway is underscored by in vitro studies showing that cultured macrophages will accumulate cholesteryl ester-rich lipid droplets when exposed to modified LDL particles that interact with scavenger receptors, and this accumulation can be abolished by including HDL in the incubation medium (Brown et al. 1980).

The HDL-mediated removal of excess cholesterol from cells occurs by a complex biochemical pathway involving multiple cell proteins and specialized transport carriers (Fig. 13.1). Studies with cultured cells have shown that when cells are exposed to cholesterol-rich medium in the absence of HDL particles, much of the excess cholesterol taken up by cells accumulates within intracellular compartments, provided cells are quiescent and do not require cholesterol for ongoing membrane synthesis (Slotte et al. 1987; Aviram et al. 1989; Oram et al. 1991). Most of this excess cholesterol is esterified by the microsomal enzyme acyl CoA cholesterol acyltransferase (ACAT) and is stored as intracellular lipid droplets. However, even if esterification of free cholesterol is blocked by an inhibitor of ACAT, much of the excess unesterified cholesterol is stored within intracellular compartments rather than transported to the plasma membrane. Morphological studies suggest that excess free cholesterol is sequestered in unique membrane-rich lamellar bodies that are formed within the cytoplasm when cells are loaded with cholesterol (McGookey and Anderson 1983). This intracellular storage mechanism may have developed to prevent over-accumulation of cholesterol within the plasma membrane where alterations in fluidity may have deleterious effects on the cell.

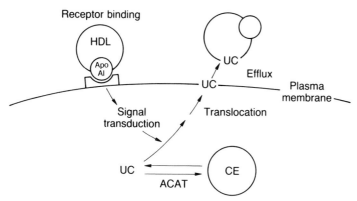

Fig. 13.1. Model for the HDL receptor-mediated transport of cholesterol from cells. Abbreviations: UC, unesterified chyolesterol; CE, cholesteryl esters; ACAT, acyl CoA cholesterol acyltransferase.

When quiescent cholesterol-loaded cells are exposed to HDL, translocation of intracellular unesterified cholesterol to the plasma membrane is stimulated (Slotte et al. 1987; Aviram et al. 1989; Oram et al. 1991). A fraction of this translocated cholesterol desorbs from the cell and is sequestered by HDL particles. This stimulatory action of HDL is mediated by reversible binding of high-density apolipoproteins to high-affinity binding sites on the cell surface (Oram et al. 1987). Apolipoprotein binding appears to transduce intracellular signals involving protein kinase C and perhaps other kinases that in turn activate the cholesterol translocation pathway, probably through phosphorylation of key transport proteins (Mendez et al. 1989, 1990).

This model assumes that several specific cell proteins are involved in the pathway for HDL-mediated cholesterol efflux, particularly a cell-surface receptor that interacts with high-density apolipoproteins and transduces appropriate signals. Thus the putative HDL receptor molecule may function to communicate to cells that exogenous acceptors for cholesterol are present and that excess cholesterol should be transported to plasma membrane domains for removal. By shunting cholesterol from intracellular storage sites to the plasma membrane, this pathway would prevent over-accumulation of intracellular cholesterol when cells are actively internalizing sterol-rich lipoproteins. Although studies to test this hypothesis are still in progress, the following sections describe some of the work from our laboratory that provided the evidence for this model of the HDL receptor pathway.

Methods

To test the hypothesis that receptor binding of HDL promotes removal of intracellular sterol, it was necessary to devise procedures to selectively

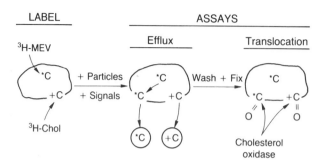

Fig. 13.2. Protocol used to study sterol efflux and translocation between intracellular pools and the media. Abbreviations: ^{3}H-MEV, [^{3}H]mevalonolactone; ^{3}H-chol, [^{3}H]cholesterol; *C, radiolabeled cholesterol.

radiolabel intracellular and plasma membrane pools of sterol in cholesterol-loaded cells and then to monitor movement of radiolabeled sterol between cellular pools and the medium. The strategy we used is shown schematically in Fig. 13.2. To radiolabel intracellular pools, cells were pulsed at 37°C for 3–4 hours with medium containing the biosynthetic precursor [^{3}H]mevalonolactone. When cells are growth arrested and loaded with cholesterol prior to the pulse incubations, the rate of biosynthesis is markedly suppressed. However, since mevalonolactone enters the biosynthetic pathway at a step beyond the enzymes that are repressed the most by cholesterol loading, trace amounts of radiolabeled sterol are synthesized in microsomal membranes. Evidence that this tracer readily enters the substrate pool for ACAT was provided by studies showing that over 70% of the newly synthesized labeled sterol is esterified (Oram et al. 1991). Thus this procedure selectively labels the intracellular pools of sterol that are accessible to ACAT in cholesterol-loaded cells. To selectively radiolabel plasma membrane cholesterol, cells were pulsed with trace quantities of [^{3}H]cholesterol (Fig. 13.2). With this protocol isotope is incorporated into the plasma membrane, but it is transported slowly into the cellular compartments.

After these pulse-labeling procedures, cells were chased at various times with medium containing lipoproteins and other possible affectors (Fig. 13.2). Cells were then chilled on ice, and the medium was collected to measure [^{3}H]sterol efflux. To monitor movement of intracellular [^{3}H]sterol between cellular pools, cells were treated with cholesterol oxidase by a modification of the method described by Lange and Ramos (1983). Since this enzyme presumably does not penetrate the plasma membrane, [^{3}H]sterol that is converted to [^{3}H]cholestonone by cholesterol oxidase treatment was assumed to represent plasma membrane-associated [^{3}H]sterol, whereas [^{3}H]sterol resistant to this treatment was assumed to be within intracellular pools. To ascertain that net movement of sterol out of microsomal pools had occurred, cells were assayed for changes in the activity of two biochemical processes that are regulated reciprocally by the level of microsomal cholesterol: rates of sterol esterification (ACAT activity) and sterol biosyntheses (Oram 1989).

High-Density Apolipoproteins Stimulate Translocation and Efflux of Intracellular Sterol

In proliferating cells that are not overloaded with cholesterol, newly synthesized tracer sterol moves rapidly to the plasma membrane (Oram et al. 1991), presumably because synthesis of the plasma membrane creates the greatest demand for cholesterol as a structural component. However, when cells are growth arrested and overloaded with cholesterol in the absence of HDL, the newly synthesized sterol tracer enters an intracellular pool that moves slowly to the plasma membrane (Slotte et al. 1987; Oram et al. 1991). This is not simply due to esterification of the sterol by ACAT, since trapping of tracer within intracellular pools occurs even in the presence of an ACAT inhibitor. Thus with reduced demand for membrane cholesterol and overabundance of available cholesterol, cells store the excess cholesterol within intracellular compartments that are not directly accessible to extracellular macromolecules.

When these cells are exposed to HDL_3, there is a rapid translocation of radiolabeled sterol from intracellular pools to the plasma membrane and an increase in sterol efflux. This stimulatory action of HDL_3 is mediated by apolipoproteins, as modification of HDL_3 by treatment with either tetranitro-methane (Slotte et al. 1987; Aviram et al. 1989) or trypsin (Oram et al. 1991) reduces its ability to stimulate intracellular sterol translocation and efflux. When plasma membrane cholesterol is radiolabeled directly by incubating cells with trace amounts of $[^3H]$cholesterol, the modified HDL_3 particles are equally as effective as native HDL_3 in promoting efflux of radiolabeled cholesterol (Slotte et al. 1987; Oram et al. 1991). Thus the interaction of HDL_3 apolipoproteins with the cell surface specifically stimulates translocation and efflux of intracellular sterol, and this effect is independent of the ability of the particle to act as an acceptor for plasma membrane cholesterol.

The stimulation of intracellular sterol translocation and efflux by high-density apolipoproteins results in a depletion of intracellular pools of cholesterol mass. Exposure of cholesterol-loaded cells to native HDL_3 but not to apolipoprotein-modified particles leads to a a suppression of ACAT activity and a reduction in cell cholesteryl ester content (Brinton et al. 1986; Oram et al. 1991). This is not because of a direct effect on ACAT, since it was associated with an increase in the rate of sterol biosynthesis from acetate. Thus the size of the intracellular pool of cholesterol that is esterified by ACAT and that down-regulates sterol biosynthesis is depleted in the presence of high-density apolipoproteins.

Intracellular Signals in High-Density Apolipoprotein-Mediated Sterol Translocation and Efflux

Results showing that high-density apolipoproteins stimulate movement of intracellular sterol under conditions where there is no detectable internalization of HDL particles (Oram et al. 1987) suggest that their effect may be mediated by intracellular signals. Initial evidence for this was provided by studies showing that incubation of cholesterol-loaded cells with HDL_3 activates protein kinase C, an enzyme commonly involved in signal transduction pathways. To test the possible involvement of protein kinase C in HDL-mediated sterol trafficking, we biosynthetically labeled intercellular pools of cholesterol-loaded cells and measured sterol efflux in the presence of the protein kinase C inhibitor sphingosine. This inhibitor reduced the ability of HDL_3 to remove intracellualr sterol to values seen in the presence of trypsin-modified HDL (Mendez et al. 1989). Other studies indicated that sphingosine suppressed the ability of HDL_3 to stimulate translocation of sterol from cholesterol oxidase inaccessible (intracellular) to oxidase-accessible (cell-surface) pools. Sphingosine had no effect on either HDL_3-mediated efflux of [^{3}H]cholesterol associated with the plasma membrane or cell-surface binding of HDL_3. Moreover, ceramide, the acylated from of sphingosine that does not inhibit protein kinase C, did not inhibit the HDL_3-stimulated translocation and efflux of intracellular sterol. These studies support the concept that the increased sterol movement associated with the interaction of HDL_3 apolipoproteins with the cell surface involves activation of protein kinase C. Additional support was obtained from studies showing that diacylglycerol and phorbol esters, which activate protein kinase C, stimulate sterol translocation from intracellular pools to the plasma membrane in the absence of HDL particles (Mendez et al. in press).

The High-Density Lipoprotein Receptor

The most straightforward explanation for the results described above is that high-density apolipoproteins interact with specific cell-surface receptors that transduce signals to intracellular targets involved in transporting cholesterol to the plasma membrane. Evidence for the existence of HDL receptors was first obtained from cell culture studies showing the existence of high-affinity binding sites on the cell surface that interact specifically with high-density apolipo-proteins. The number of binding sites increases when cells are either loaded with cholesterol (Oram et al. 1983) or growth arrested (Oppenheimer et al. 1988). Thus, the same culture conditions that promote trapping of excess cholesterol in intracellular pools also increase the number of cell-surface binding sites for HDL, consistent with the idea that these binding sites function to rid cells of excess cholesterol. Ligand blot studies revealed that cell membranes contain a 110-kDa protein that preferentially binds high-density apolipoproteins and that undergoes up-regulation in response to cholesterol loading of cells (Graham and Oram 1987) or inhibition of cell proliferation (Oppenheimer et al. 1988). Cell fractionation studies revealed that this protein is associated with the plasma membrane (Hokland et al. 1990). Thus, this protein represents the best candidate for an HDL receptor yet identified. A cDNA clone for a candidiate receptor protein has also been identified in our laboratory. Studies are in progress to characterize these proteins and to determine if they represent the same putative HDL receptor molecule.

Clinical Implications

The identification of an HDL receptor pathway that facilitates removal of excess cholesterol from cells has obvious relevance to the disease of atherosclerosis. This complex cellular pathway is comprised of multiple proteins, any one of which may be defective in genetic disorders yet to be discovered. Since a higher incidence and prevalence of atherosclerosis is frequently associated with low levels of HDL cholesterol, even within groups with relatively normal levels of LDL cholesterol, it is attractive to speculate that low HDL cholesterol may in some cases reflect a defective HDL receptor pathway. The HDL receptor hypothesis also raises the possibility that specific pharmacological agents could be developed that act at the cellular level to retard progression or enhance regression of atherosclerosis, even in subjects with normal HDL receptor pathways. For example, an agonist for the HDL receptor may enhance the rate of clearance of cholesterol from arterial cells when faced with an excessive rate of uptake of LDL cholesterol. Clearly, an understanding of the properties of the HDL receptor pathway and its protein components will provide insight into possible cellular mechanisms that underlie some forms of atherosclerosis and suggest future therapy.

References

Aviram M, Bierman EL, Oram JF (1989) High density lipoprotein stimulates sterol translocation between intracellular and plasma membrane pools in human monocyte-derived macrophages. J Lipid Res 30:65–76

Brinton EA, Oram JF, Chen CH, Albers JJ, Bierman EL (1986) Binding of high density lipoproteins to cultured fibroblasts after chemical alteration of apoprotein amino acid residues. J Biol Chem 261:495–503

Brown MS, Goldstein JL (1983) Lipoprotein metabolism in the macrophage. Ann Rev Biochem 52:223–261

Brown MS, Ho YK, Goldstein JL (1980) The cholesterol ester cycle in macrophage foam cells: continual hydrolysis and reesterification of cytoplasmic cholesterol esters. J Biol Chem 255:9344–9452

Glomset JA (1968) The plasma lecithin:cholesterol acyltransferase reaction. J Lipid Res 9:155–167

Graham DL, Oram JF (1987) Identification and characterization of a high density lipoprotein-binding protein in cell membranes by ligand blotting. J Biol Chem 262:7439–7442

Hokland BM, Mendiz AJ, Oram JF (1990) Subcellular localization and tissue distribution of the 110 kDa HDL receptor protein. Arteriosclerosis 10:767a

Lange Y, Ramos BV (1983) Analysis of the distribution of cholesterol in the intact cell. J Biol Chem 258:15130–15134

McGookey PJ, Anderson RGW (1983) Morphological characterization of the cholesteryl ester cycle in cultured mouse macrophage foam cells. J Cell Biol 97:1156–1168

Mendez AJ, Oram JF, Bierman EL (1989) Sphingosine inhibits HDL-mediated efflux of intracellular sterols. Atherosclerosis 9:720a

Mendez AJ, Oram JF, Bierman EL (1990) HDL binding to its receptor stimulates diacylglycerol formation in cholesterol-loaded fibroblasts. Arteriosclerosis 10:768a

Mendez AJ et al. (1991, in press) J Biol Chem

Oppenheimer MJ, Oram JF, Bierman EL (1988) Up-regulation of high density lipoprotein receptor activity by interferon associated with inhibition of cell proliferation. J Biol Chem 263:19318–19323

Oram JF (1989) Receptor-mediated transport of cholesterol between cultured cells and high-density lipoproteins. Methods Enzymol 30:65–76

Oram JF, Brinton EA, Bierman EL (1983) Regulation of high density lipoprotein receptor activity in cultured human skin fibroblasts and human arterial smooth muscle cells. J Clin Invest 721:611–1621

Oram JF, Johnson CJ, Brown TA (1987) Interaction of high density lipoprotein with its receptor on cultured fibroblasts and macrophages: evidence for reversible binding at the cell surface without internalization. J Biol Chem 262:2405–2410

Oram JF, Mendez AJ, Slotte PS, Johnson TF (1991) HDL apolipoproteins mediate removal of sterol from intracellular pools but not from plasma membranes of cholesterol-loaded fibroblasts. Arteriosclerosis 11:403–414

Slotte JP, Oram JF, Bierman EL (1987) Binding of high density lipoprotein to cell receptors promotes translocation of cholesterol from intracellular membranes to the cell surface. J Biol Chem 262:12904–12907

Steinberg D, Parthasarathy S, Carew TB et al. (1989) Beyond cholesterol. N Engl J Med 320:915–924

Chapter 14

High-Density Lipoprotein Cholesterol, Triglycerides and Coronary Heart Disease

J. R. Patsch and G. Miesenboeck

High-density lipoprotein (HDL) cholesterol, plasma levels of the triglyceride (TG)-rich lipoproteins and the risk of coronary heart disease (CHD) are interrelated. There exists a strong negative association between levels of HDL cholesterol and CHD risk, a weaker positive relation between TGs and CHD risk, and a strong inverse association between plasma levels of HDL and of the TG-rich lipoproteins. When HDL cholesterol TGs and CHD risk are considered *together*, the status of each in the triad becomes less clear.

There are two major hypotheses about the role of HDL in the development of CHD. One assigns HDL a causally protective effect against atherosclerosis. In this scenario, HDL particles trap excess cholesterol from cellular membranes by esterification and transfer the esterified cholesterol to TG-rich lipoproteins, which are subsequently removed by hepatic receptors. This reverse cholesterol transport from peripheral cells to the liver counteracts the deposition of cholesterol at sites where an excessive cholesterol load produces atherosclerosis. Thus HDL cholesterol levels would signify a high rate of reverse cholesterol transport. According to the second hypothesis HDL particles do not have a causal role: they do not interfere directly with cholesterol deposition in the arterial wall but rather reflect the metabolism of TG-rich lipoproteins and their conversion to atherogenic remnants. In this view HDL cholesterol levels indicate an efficient metabolism of TG-rich lipoproteins and a low production rate of atherogenic remnants.

Several major prospective epidemiologic studies have demonstrated a strong inverse, independent relation between HDL cholesterol and CHD risk. Gordon et al. (1989) reanalyzed data from the Framingham Heart Study (FHS), the Lipid Research Clinics Prevalence Mortality Follow-up Study (LRCF) and the control groups of the Coronary Primary Prevention Trial (CPPT) and the Multiple Risk Factor Intervention Trail (MRFIT), and found this strong and independent relation to stand. They concluded that a 1-mg/dl increment in HDL cholesterol is associated with a CHD risk decrement of 2% in men and 3% in women (see also Gordon and Rifkind 1989).

In contrast to HDL cholesterol, TGs are not generally accepted as a risk factor for CHD. While an association of elevated TG levels with CHD is usually shown on univariate analyses the relationship tends to break down when multiple factors are taken into account. The factor that eliminates TGs on these analyses is HDL cholesterol (Austin 1989). This fact, however, does not necessarily disqualify TGs or TG-rich lipoproteins as causative agents for CHD. Over a wide range, HDL cholesterol levels are determined largely by the metabolism of TG-rich lipoproteins. The influence of TG-rich lipoproteins on HDL is particularly pronounced in the postprandial state with the entry of chylomicrons into the circulation. Therefore, HDL cholesterol can be viewed as an integrative marker of TG transport in all states of absorption that tends to eliminate the rapidly fluctuating TGs as a risk factor (Miesenböck and Patsch 1990).

It is not the mature, spherical HDL particles that are secreted into the circulation but discoidal precursor particles devoid of cholesteryl esters. The action of the cholesterol esterifying plasma enzyme lecithin:cholesterol acyltransferase (LCAT) transforms the nascent HDLs into mature particles. There are two major subfractions of spherical HDL: the small, lipid-poor and dense HDL_3 and the larger more lipid-rich and less dense HDL_2. HDL_3 plasma levels are fairly constant in an individual, whereas HDL_2 levels vary greatly and thus account for most of the variability of total HDL cholesterol. Thus the more interesting HDL subfraction in terms of CHD risk assessment is HDL_2.

TG-rich lipoproteins affect HDL_2 levels in two ways. First, the rapid lipolysis of TG-rich lipoproteins by the enzyme lipoprotein lipase (LPL) generates excess surface material (predominantly phospholipids), which is assimilated by HDL. This uptake of lipid promotes the formation of the large HDL_2 and thereby raises the cholesterol-carrying capacity of HDL. The assimilation of lipolytic surface remnants by HDL not only effects the formation of HDL_2 but also may protect arterial wall cells: surface remnants when not incorporated into HDL are cytotoxic to macrophages in culture (Chung et al. 1989). Second, delayed lipolysis of TG-rich lipoproteins increases the opportunity for the reciprocal transfer of TGs from TG-rich lipoproteins into HDL particles and of cholesteryl esters from HDL into TG-rich lipoproteins. This neutral lipid exchange reaction is catalyzed by lipid transfer protein I (LTP-I) (Albers et al. 1984; Tall 1986). TGs transferred to HDL are hydrolyzed by the enzyme hepatic lipase (Patsch et al. 1984). In this way, HDL cholesterol is lost in TG-rich lipoproteins, large HDL particles are converted into small HDL particles, and the cholesterol-carrying capacity of HDL decreases. Hence, rapid lipolysis of TG-rich lipoproteins keeps HDL_2 levels (and thus HDL cholesterol) high by promoting the formation of the larger HDL_2 particles as well as by preventing their catabolism. The response of HDL cholesterol to the rapid fluctuations of the concentrations of TG-rich lipoproteins is comparably slow and truncated and constitutes the biochemical basis for the "memory" of HDL with respect to TG transport (Miesenböck and Patsch 1990).

Lipolysis as a determinant of HDL_2 levels may be the mechanism linking low HDL cholesterol to the cardiovascular risk constellation of obesity, hyperinsulinemia and insulin resistance. In 146 healthy subjects of either sex, 41% of the HDL_2 variance was explained by the combined effect of the waist-to-hip ratio as a measure of truncal obesity, the plasma insulin level and the degree of glucose intolerance (Ostlund et al. 1990). In a different study, weight reduction and maintenance of a reduced-obesity state lowered fasting TG levels, increased HDL cholesterol, and raised the concentration of HDL_2. The HDL_2:HDL_3 cholesterol ratio was strongly correlated to the insulin responsiveness of adipose tissue LPL but not of post-heparin plasma LPL. This suggests a tissue-specific regulation of LPL – and of HDL_2 levels – by insulin (Eckel and Yost 1989). Both studies indicate that insulin resistance limits LPL action of adipose tissue and thereby decreases HDL_2.

HDL levels are determined not only by the metabolism of TG-rich lipoproteins but also by the secretion of HDL precursors. Overexpression of human apolipoprotein (apo) A-I in transgenic mice greatly augmented the concentration of this apolipoprotein in the animals' plasma. The apo A-I pool size was directly related to HDL cholesterol levels. The slope of the regression line indicated that the apo A-I/cholesterol mass ratio of the particles was 0.26, equaling that of human HDL_3 (Walsh et al. 1989). Provided that apo A-I is secreted in association with phospholipids, the enhanced biosynthetic rate of apo A-I can be equated with an increased availability of HDL precursors. In this experimental situation plasma apo A-I levels correlated with HDL cholesterol. However, the elevation was confined to small particles such as HDL_3, whereas the concentration of HDL_2 remained unaffected (Walsh et al. 1989). This result underscores that another factor, presumably the supply of surface remnants released from TG-rich lipoproteins through lipolysis, is rate limiting for the formation of HDL_2.

According to the above-described causal hypothesis, one essential component of reverse cholesterol transport is the transfer of cholesteryl esters from HDL particles to TG-rich lipoproteins. In this way, cholesterol transported with HDL can enter the final leg in its tour from peripheral cells to the liver. In the non-causality scenario, however, the LTP-I reaction has a role diametrically opposite to its assigned part in reverse cholesterol transport. A loss of HDL cholesterol into TG-rich lipoproteins is considered potentially harmful since cholesterol would be redistributed from the system of antiatherogenic particles into the one of atherogenic particles. LTP-I would thus provide a mechanism for switching "good" cholesterol into "bad" cholesterol. Therefore, absence of LTP-I should be a decisive test for the two metabolic models under discussion. If the reverse cholesterol transport hypothesis is correct, the deficiency should disrupt this pathway and thus *predispose* to CHD. Alternatively, if HDL is an indicator for TG transport, LTP-I deficiency should prevent the additional enrichment of TG-rich lipoproteins with HDL-derived cholesterol and thus *protect* against CHD.

Indeed a genetic LTP-I deficiency has been described in two Japanese siblings (Koizumi et al. 1985). The disorder is caused by substitution of adenine for guanine at the 5' splice donor site of intron 14 of the LTP-I gene (Brown et al. 1989). Because of the absence of LTP-I, HDL particles fail to become enriched in TGs, which prevents their degradation by hepatic lipase. The result is extremely elevated HDL cholesterol values (248 and 175 mg/dl in the Japanese siblings) and an overwhelming preponderance of the large-HDL subfraction. Plasma levels of low-density lipoprotein (LDL) are decreased, presumably because the reciprocal transfer of HDL cholesteryl esters into chylomicrons and very-low-density lipoproteins (VLDLs) – and ultimately into LDLs – is blocked also. LTP-I deficiency segregates as a familial trait of longevity and resistance to caridovascular disorders. This experiment of nature argues strongly against the causal hypothesis of reverse cholesterol transport.

More recently, attempts have been made to delineate the influence of neutral lipid transfer on the reaction partners of HDL, the TG-rich lipoproteins. Monoclonal-antibody inhibition of LTP-I in the rabbit produced the expected compositional changes of both lipoprotein reaction partners. HDL cholesteryl esters rose significantly, with a concomitant fall in HDL TGs, and the cholesteryl ester-to-triglyceride ratio of VLDLs declined significantly (Whitlock et al. 1989). This result provides direct evidence that part of the cholesteryl esters transported with TG-rich lipoproteins originate in HDL. In a study in man, the effect of postprandial lipemia on the distribution of cholesteryl esters in plasma lipoproteins was investigated (Dullaart et al. 1989). In going from the postabsorptive to the postprandial state, the fraction of cholesteryl esters transported in TG-rich lipoproteins rose 25-fold compared with 15% decreases in those fractions in LDL and HDL. Whereas the transfer of cholesteryl esters from HDL particles to TG-rich lipoproteins leaves these cholesteryl esters in the same metabolic cascade, the cholesteryl esters transferred from HDL particles to TG-rich lipoproteins are translocated from the pool of "good" to "bad" cholesterol.

The biochemical and clinical data regarding HDL cholesterol and CHD reviewed in this chapter support the non-causality concept that the negative association between HDL cholesterol and CHD does not depend solely on a direct antiatherogenic action of HDL. The non-causality concept, however, cannot be interpreted to suggest that HDL particles are innocent bystanders that are only passively affected by the truly atherogenic events. On the contrary, the LTP-I mechanism for switching "good" cholesterol to "bad" cholesterol assigns HDL a central role in the pathogenesis of atherosclerosis. The driving force for the switch, however, is the metabolism of TG-rich lipoproteins: rapid clearance of TG-rich lipoproteins promotes the formation of HDL_2, and low levels of TG-rich lipoproteins do not allow excessive transfer of HDL cholesteryl esters into TG-rich lipoproteins. This keeps the cholesterol-carrying capacity of HDL – i.e., HDL cholesterol – high and the net effect is antiatherogenic. Delayed clearance and accumulation of TG-rich lipoproteins do not allow the formation of HDL_2 and lead to excessive loss of cholesteryl esters from HDL into lipoprotein fractions associated with high CHD risk. HDL cholesterol is reduced and the net effect is atherogenic.

References

Albers JJ, Tollerson JH, Chen C-H, Steinmetz A (1984) Isolation and characterization of human plasma lipid transfer proteins. Arteriosclerosis 4:49–58

Austin ME (1989) Plasma triglyceride as a risk factor for coronary heart disease: the epidemiologic evidence and beyond. Am J Epidemiol 129:249–259

Brown ML, Inazu A, Hesler CE et al. (1989) Molecular basis of lipid transfer protein deficiency in a family with increased high-density lipoprotein. Nature 342:448–451

Chung BH, Segrest JP, Smith K, Griffin FM, Brouillette CG (1989) Lipolytic surface remnants of triglyceride-rich lipoproteins are cytotoxic to macrophages but not in the presence of high density lipoprotein: a possible mechanism of atherogenesis? J Clin Invest 83:1363–1374

Dullaart RP, Groener JE, van Wijk H, Gluiter WJ, Erkelens DW (1989) Hereditary lipemia-induced redistribution of cholesteryl ester between lipoproteins. Studies in normolipidemic, combined hyperlipidemic, and hypercholesterolemic men. Arteriosclerosis 9:614–622

Eckel RH, Yost TJ (1989) HDL subfractions and adipose tissue metabolism in the reduced-obese state. Am J Physiol 256:E740–E746

Gordon DJ, Rifkind BM (1989) High-density lipoprotein: the clinical implications of recent studies. N Engl J Med 321:1311–1316

Gordon DJ, Probstfield JL, Garrison RJ et al. (1989) High-density lipoprotein cholesterol and cardiovascular disease: four prospective American studies. Circulation 79:8–15

Koizumi J, Mabuchi H, Yoshimura A et al. (1985) Deficiency of serum cholesteryl-ester transfer activity in patients with familial hyperalphalipoproteinemia. Atherosclerosis 58:175–186

Miesenböck G, Patsch JR (1990) Relationship of triglyceride and high-density lipoprotein metabolism. Atherosclerosis Rev 18:119–128

Ostlund RE Jr, Staten M, Kohrt WM, Schultz J, Malley M (1990) The ratio of waist-to-hip circumference, plasma insulin level, and glucose intolerance as independent predictors of the HDL_2 cholesterol level in older adults. N Engl J Med 322:229–234

Patsch JR, Prasad S, Gotto AM Jr, Bengtsson-Olivecrona G (1984) Postprandial lipemia: a key for the conversion of high-density $lipoprotein_2$ into high-density $lipoprotein_3$ by hepatic lipase. J Clin Invest 74:2017–2023

Tall Ar (1986) Plasma lipid transfer proteins. J Lipid Res 27:361–367

Walsh A, Ito Y, Breslow JL (1989) High levels of human apolipoprotein A-I in transgenic mice result in increased plasma levels of small high density lipoproteins (HDL) comparable to human HDL_3. J Biol Chem 264:6488–6494

Whitlock ME, Swenson TL, Ramakrishnan R et al. (1989) Monoclonal antibody inhibition of cholesteryl ester transfer protein activity in the rabbit. J Clin Invest 84:129–137

Chapter 15

Apolipoproteins, Reverse Cholesterol Transport and Coronary Heart Disease

G. Assmann, A. von Eckardstein and H. Funke

Introduction

Several epidemiological and clinical studies revealed an inverse correlation between low plasma concentrations of high-density lipoprotein (HDL) cholesterol as well as its major protein component apolipoprotein A-I (apo A-I) and the risk of myocardial infarction (reviewed in Gordon and Rifkind 1989). Family and twin studies suggested partial heredity of low HDL cholesterol levels and have put the influence of genes at 35%–50% (Hunt et al. 1989; Assmann et al. 1989a). Frequently, familial HDL cholesterol deficiency was paralleled with a family history of premature coronary heart disease (CHD) (Pometta et al. 1979; DeBacker et al. 1986). Thus, a causal role between low levels of HDL cholesterol and coronary risk appears well established. However, the pathophysiological relationship between decreased serum concentrations of HDL cholesterol and the development of coronary heart disease has still remained obscure. The reverse cholesterol transport model (Glomset 1968) is most widely used to explain the role of HDL in lipid metabolism and in atherogenesis. HDL precursors, so-called nascent HDL or HDL disks, are generated through lipolysis of chylomicrons and VLDL as well as by direct secretion of the liver (reivewed in Brunzell 1989 and Assmann et al. 1989b). HDL subclasses and reconstituted HDL-like particles that contain apo A-I were shown to take up excess cellular cholesterol by interaction with specific cell surface recognition sites (reviewed in Assmann et al. 1989b). Upon esterification of free cholesterol by lecithin:cholesterol acyltransferase (LCAT), the core of small and dense HDL particles becomes enriched in cholesterol esters and their particle size increases (reviewed in Norum et al. 1989). At least four different metabolic pathways have been identified, by which cholesterol esters can be delivered to the liver, where they can be disposed of via bile acid synthesis (reviewed in Assmann et al. 1989b): (1) a subpopulation of HDL acquires apo E and can be recognized by hepatic apo E receptors; (2) HDL without apo E can also be endocytosed by hepatocytes; (3) hepatic triglyceride lipase (HTGL) mediates the uptake of cholesterol esters into liver cells; (4) cholesterol ester transfer protein (CETP) catalyzes the net transfer of HDL cholesterol esters to LDL, IDL and VLDL which are taken up via hepatic apo B, E receptors. Genetic defects that interfere with the regular structure of HDL or with processes important for the generation or removal of these lipoproteins may be disadvantageous for their carriers' health. We used different methodological approaches to identify genetic causes for HDL deficiency and to understand their pathophysiological role: (1) association and cosegregation studies using restriction fragment length polymorphisms (RFLPs) and haplotypes of candidate genes; (2) screening for structural apolipoprotein mutants; and (3) sequencing of candidate genes of individuals suffering from severe HDL deficiency syndromes.

The Role of the Apo A1/C3/A4 Gene in Association Studies and Cosegregation Studies

Eight dimorphic sites in the APOLP1 locus containing the genes for apolipoproteins A-I, C-III and A-IV were determined by RFLP analysis in 314 survivors of myocardial infarction and 267 medical students as controls. All but one single RFLP sites were found at identical frequencies in both groups. Only the X2 allele (XmnI site 5' of apo A1) was observed with significantly different frequencies ($p < 0.05$, Student t-test). Genotypes obtained from the eight sites were distributed at similar frequencies in both populations and none of the more frequent APOLP1 genotype appears to be associated with increased coronary risk in the population (Assmann et al. 1989a). Moreover, cosegregation studies in seven of eight families with vertically transmitted reduction of HDL cholesterol serum concentrations excluded a causative role of the APOLP1 locus (Assmann et al. 1989a). In summary, this locus either does not contribute much to HDL deficiency in the general population or the defects are multiallelic. The identification of various apo A-I mutants leading to HDL deficiency (see below) therefore may be a consequence of bias, because much interest has been paid to this locus. However, defects at this locus could also have evaded detection by association as well as cosegregation studies, because of the high frequency of the more common alleles. Furthermore, association studies fail to detect multiallelic defects and cosegregation studies suffer from the variance of phenotype expression in HDL-deficient families.

The Role of Apo A-I Mutants

In the course of an elecrophoretic mutation screening program of 32 000 dried blood samples from newborns, 16 genetic variants of apo A-I were found by the presence of additional bands upon isoelectric focusing (IEF). Structural analyses of the mutant proteins by the combined use of high-performance liquid chromatography, time-of-flight secondary ion mass spectrometry and sequence analysis led to the identification of the following defects: $Pro_3 \rightarrow Arg$ (1×), $Pro_4 \rightarrow Arg$ (1×), $Lys_{107} \rightarrow 0$ (4×), $Lys_{107} \rightarrow Met$ (2×), $Glu_{139} \rightarrow Gly$ (2×), $Glu_{147} \rightarrow Val$ (1×), $Pro_{165} \rightarrow Arg$ (4×) and $Glu_{198} \rightarrow Lys$ (1×) (von Eckardstein et al. 1989, 1990). All but one of these apo A-I variants were found associated with normal HDL cholesterol and apo A-I plasma concentrations: studies in three unrelated families with apo A-I($Pro_{165} \rightarrow Arg$) revealed that subjects heterozygous for this mutant ($n=12$) exhibit lower mean values for apo A-I (109 ± 16 mg/dl) and HDL cholesterol (37 ± 9 mg/dl) than unaffected family members ($n=9$): 176 ± 41 mg/dl and 64 ± 18 mg/dl, respectively ($p<0.001$). In 9 of 12 apo A-I($Pro_{165} \rightarrow Arg$) carriers, the concentrations of apo A-I were below the 5th percentile of sex-matched controls. Using two-dimensional immunoelectrophoresis and densitometry, the relative concentration of the variant apo A-I in heterozygous carriers of apo A-I($Pro_{165} \rightarrow Arg$) was determined to account for only 30% of the total plasma apo A-I mass instead of the expected 50%. Thus, the observed apo A-I deficiency may largely be a consequence of the decreased concentration of the variant apo A-I. In reconstituted HDL, apo A-I($Pro_{165} \rightarrow Arg$) exhibited binding activities towards POPC and DPPC identical to normal apo A-I (Jonas et al. 1991). LCAT cofactor activity of HDL reconstituted with apo A-I($Pro_{165} \rightarrow Arg$) was 30% lower than HDL reconstituted with normal apo A-I from the same donor. However, the variance between several normal apo A-I preparations was higher than this deviation (Jonas et al. 1991). Thus, the question is unanswered whether or not decreased LCAT cofactor activity of apo A-I($Pro_{165} \rightarrow Arg$) is relevant for the pathogenesis of HDL cholesterol deficiency in affected individuals.

To date, structural analysis of 37 instances of apo A-I variants found by electrophoretic screening of approximately 45 000 samples or in diseased persons led to the identification of 25 different structural variations in apo A-I (reviewed in Assmann et al. 1990). Only a minority of apo A-I variants was found associated with lipid disorders. Some of them did not appear of vital disadvantage: apo A-I($Arg_{173} \rightarrow Cys$), apo A-I($Pro_{165} \rightarrow Arg$) and apo A-I(202 frameshift) (all reviewed in Assmann et al. 1990). However there have also been reports on three forms of apo A-I deficiency that were associated with premature myocardial infarction (reviewed in Breslow 1989 and Assmann et al. 1990) and heterozygosity for apo A-I($Arg_{26} \rightarrow Gly$) (Nichols et al. 1988), which is associated with lethal familial amyloidotic polyneuropathy. In order to analyze if regions within apo A-I were affected by genetic variation to a different degree, we statistically evaluated the distribution of amino acid substitutions throughout the apo A-I primary sequence (von Eckardstein et al. 1990). This analysis revealed a significant over representation of "electrically non-neutral" amino acid substitutions in the amino terminal portion of apo A-I (residues $1 \rightarrow 10$: $p<0.001$) and in large part of the protein's α-helical domain (residues $103 \rightarrow 198$: $p<0.05$; residues $103 \rightarrow 177$: $p<0.025$). Comparison of mutation sites in the human apo A-I gene with sites of non-synonymous substitutions observed by interspecies comparisons of apo A-I sequences from man, rat, pig, monkey and dog revealed that amino acid substitutions found in human apo A-I were predominantly localized in areas which were little conserved during mammalian evolution. These regions may therefore represent areas of less structural constraint for the function of apo A-I. Theoretically, amino acid substitutions in stringent functional domains of apo A-I can result in apo A-I deficiency and may therefore escape detection procedures (von Eckardstein et al. 1990). Amino acid substitutions affecting the amphipathic α-helices of apo A-I (residues 99–186) were predominantly located on their hydrophilic site. Due to the limited sensitivity of isoelectric focusing in carrier ampholyte containing pH gradients, electrically neutral amino acid substitutions escape detection. Statistically, the majority of amino acid substitutions affecting the hydrophobic sites of the amphipathic α-helices are electrically neutral. The hydrophobic sites probably are relevant for the mediation of lipid binding and LCAT activating properties of apo A-I. Therefore, further screening for apo A-I variants should apply techniques that are sensitive for deficiency variants as well as electrically neutral amino acid substitutions.

Molecular Defects in HDL Cholesterol Deficiency Syndromes

Another approach to the identification of genes regulating HDL cholesterol serum concentration is the elucidation of the molecular bases in severe HDL deficiency syndromes. Recently, we identified mutations leading to HDL cholesterol deficiency syndromes, in which homozygous patients are characterized by massive corneal opacifications and decreased plasma activity for esterification of cholesterol (Table 15.1): fish-eye disease (FED) (Funke et al. 1991), apo A-I deficiency with corneal opacifications and LCAT deficiency (Funke et al. 1991).

Table 15.1. Clinical and biochemical findings in LCAT deficiency

	LCAT deficiency	FED	Apo A-I(202 frameshift)
Affected gene	LCAT	LCAT	Apo A-I
Major clinical findings	Anemia Proteinuria Uremia	Visual impairment	Visual impairment
Corneal opacities	+++	+++	+++
Serum cholesterol	Normal/high	Normal	Normal
Serum triglycerides	Normal/high	High	Increased
Cholesterol esters	Low	Low normal	Low normal
LDL cholesterol	Normal/high	Normal	Normal
HDL cholesterol	0	0	0
Apo A-I	40%	15%	2%
Apo A-II	30%	10%	20%
Cholesterol esterification rate[a]	0	100%	80%
LCAT activity[b]	0	0	40%
LCAT mass	Low	50%	40%
LCAT specific activity	0%–10%	0	100%

[a] The cholesterol esterification rate reflects the esterification of endogenous cholesterol and was determined following the method of Dobiasova (1983)
[b] An exogenous substrate assay using artificial proteoliposomes that contain apo A-I as well as phosphatidylcholine and unesterified [^{3}H]cholesterol was utilized to measure plasma LCAT activity (Bazri and Korn 1973).

Fish-Eye Disease

The presence of massive corneal opacifications, HDL cholesterol deficiency and the absence of LCAT in vitro activity in spite of near-normal ratios of free cholesterol/cholesterol esters and a normal in vitro cholesterol esterification rate found in two 60 and 70-year-old German brothers suggested the presence of FED (reviewed in Norum et al. 1989). Since LCAT mass was found in considerable amounts in plasma while its activity was largely reduced, FED was assumed to be caused by a defective but partially active LCAT gene product. Amplification by the polymerase chain reaction (PCR) and subsequent direct sequencing of the LCAT genes showed that the two brothers were homozygous for a C to T exchange in the second base of codon 123 which leads to a substitution of a threonine with an isoleucine. Cosegregation of this defect with decreased *specific* LCAT activity in the analyzed family (19 members) established the existence of a causal relationship between the two. Further evidence was derived from the assessment of the $Thr_{123} \rightarrow Ile$ substitution in two Dutch patients unrelated to the German FED family who also presented the clinical and biochemical sign of the FED phenotype (unpublished, referred to our laboratory by J. Kastelein, Amsterdam).

Apo A-I (202 Frameshift)

A 42-year-old patient who presented with a phenotype which closely resembled FED by the presence of massive corneal opacifications, the virtual absence of HDL cholesterol and a 50% reduction of LCAT activity in vitro was analyzed for the underlying biochemical defect. IEF and subsequent immunoblot analysis of the homozygous patient's plasma showed atypical banding for apo A-I. Direct sequencing of the patient's apo A-I genes subsequent to their PCR amplification showed a homozygous single-base deletion in codon 202 which results in the formation of a frameshift protein consisting of 229 instead of the normal 243 amino acids. The defect cosegregated with reduced serum concentrations for apo A-I and HDL cholesterol in heterozygous family members. The variant protein's residues $203 \rightarrow 229$ differ from the sequence of normal apo A-I and contain three cysteines, an amino acid that is normally absent from apo A-I. The observation of an atypical banding with more than the expected three bands upon IEF and the immunological colocalization of the mutant apo A-I with apo A-II in Western blots led to the hypothesis that the mutant apo A-I forms hetero-oligomers with other cysteine-containing proteins. This subject was more directly addressed by mass spectrometric analysis of a band in IEF gels which was immunoreactive for apo A-I and apo A-II. Masses of 8820 Da (a), 17 625 Da (b), 26 490 Da (c) and 35 315 Da (d) were found, which correspond to the masses expected for apo A-II monomer (a), apo A-II homodimer (b), mutant apo A-I (c) and apo A-II/mutant apo A-I heterodimer (d) (Funke et al. 1991).

Familial LCAT Deficiency

We also analyzed the underlying defect in five unrelated patients from Italy (Utermann et al. 1972), Canada (Frohlich et al. 1988), France, Denmark and

Norway (unpublished, referred to us by J. Dachet and B. Jacotot, Paris, U. Gerdes and O. Faergeman, Aarhus, and T. Leren and K. Berg, Oslo), who in spite of varying clinical symptoms exhibited biochemical characteristics of familial LCAT deficiency: severely decreased HDL cholesterol plasma concentrations, increased ratios of free cholesterol/cholesterol esters and inability to esterify endogenous and exogenous cholesterol in vitro. Direct sequencing of the patients' LCAT genes subsequent to PCR amplification revealed the presence of different mutations in the LCAT gene (unpublished) which cosegregated with half-normal LCAT activity in heterozygous members of the respective families. All defects were different from the one observed in the German and Dutch FED patients.

The finding of different defects in three forms of HDL cholesterol deficiency syndromes raises two major questions: (1) how do they lead to HDL deficiency? and (2) do they predispose to increased coronary risk?

1. In vitro, each of the three clincial conditions is characterized by an impairment of cholesterol esterification, but quantitatively and qualitatively to a different degree (Table 6.1). In LCAT deficiency, esterification of cholesterol is severely impaired with any substrate, and also in vivo the proportion of esterified cholesterol is severely decreased. The underlying changes in the LCAT primary structure therefore seem to disintegrate the enzyme's ability for cholesterol esterification. In FED, near-normal in vivo ratios of free cholesterol/total cholesterol and a normal esterification of endogenous cholesterol as well as of cholesterol in VLDL and LDL (Carlson and Holmquist 1985a) suggest that LCAT in FED selectively fails to esterify cholesterol in small substrate particles which otherwise are the preferred substrate of LCAT. This activity LCAT has been termed α-LCAT (Carlsson and Holmquist 1985b). In apo A-I (202 frameshift), decreased LCAT activity apparently is caused by the reduction of LCAT mass. This was similarly observed in apo A-I/C-III deficiency (Forte et al. 1984), in Tangier disease (Carlsson et al. 1987, Pritchard et al. 1988) and in apo A-I (Milano) (Franceschini et al. 1990). Possibly, the loss of LCAT mass is caused by the lack of small-size HDL particles which are preferred substrates of LCAT or, alternatively, by hypercatabolism of lipoproteins to which LCAT is attached.

2. Whereas the causal relationship between the above-described defects and reduced HDL cholesterol serum concentrations appears very obvious, the question is open whether or not these HDL cholesterol deficiency syndromes cause CHD: the carrier of the apo A-I frameshift mutant did not exhibit any symptoms of CHD and also his family history was not in support of such an association. In contrast to the Swedish and Candian FED patients, one German FED patient had suffered from angiographically assessed CHD since the age of 50. However, his identification is biased as he was referred to us by a cardiology department. Family histories of both German and Dutch FED families did not reveal any increased incidence of myocardial infarctions. Several patients affected by LCAT deficiency suffered from myocardial infarction but the question is unanswered whether coronary risk was increased because of reduced HDL cholesterol or was secondary to renal disease, which is typical for this condition. The knowledge of the above-described as well as other defects leading to HDL deficiency will enable us to screen larger populations for their presence and to answer the

question whether or not these mutations put affected individuals at increased coronary risk.

Conclusions

The basis of genetically decreased HDL cholesterol serum concentrations appears to be multigenic and multiallelic and the phenotypic expression of this condition appears to be highly variable due to many genetic and environmental factors affecting HDL metabolism. To date, only a few cases of familial HDL cholesterol deficiency have been traced back to their genetic origin: apo A-I/C-III deficiency, apo A-I/C-III/A-IV deficiency, HDL deficiency with planar xanthomas, apo A-I(202 frameshift), apo A-I($Arg_{173} \rightarrow Cys$), apo A-I($Pro_{165} \rightarrow Arg$) (all reviewed in Breslow 1989 and Assmann et al. 1990), fish-eye disease (Funke et al. 1991), some cases of LCAT deficiency (Funke et al. 1991), and LPL($Gly_{188} \rightarrow Glu$) (Wilson et al. 1990). Most of these HDL cholesterol deficiency syndromes have been identified in homozygous patients, who suffered from distinct clinical symptoms like early onset CHD, xanthomatosis, neuropathy, nephropathy or corneal opacities, and from virtual absence of HDL cholesterol. Heterozygosity of these conditions could account for a sizeable proportion of sporadic HDL cholesterol deficiency in the population, but to date they cannot be biochemically differentiated from one another. Furthermore, because only few defects decreasing HDL cholesterol have been identified and because coronary artery disease is very frequent in the population due to other prevalent risk factors, we cannot yet judge their contribution to atherogenesis. Only large case numbers for any given defect reducing HDL plasma concentrations will allow us a definite assignment of HDL cholesterol deficiency and associated coronary risk. Moreover, the knowledge of genetic defects might allow the development of more sophisticated techniques to differentiate low HDL cholesterol plasma concentrations biochemically.

References

Assmann G, Schulte H, Funke H, von Eckardstein A, Seedorf U (1989a) The prospective cardiovascular Münster (PROCAM) study: identification of high-risk individuals for myocardial infarction and the role of HDL. In: Miller NE (ed) High density lipoproteins and atherosclerosis II. Exerpta Medica, Amsterdam, pp 51–60

Assmann G, Schmitz G, Brewer HB (1989b) Familial high density lipoprotein deficiency: tangier disease. In: Scriver CR, Beaudet AL, Sly WS, Valle D (eds) The metabolic basis of inherited disease, 6th edn. McGraw-Hill Information Services, New York, pp 1267–1282

Assmann G, Schmitz G, Funke H, von Eckardstein A (1990) Apolipoprotein A-I and HDL deficiency. In: Mahley RW, Utermann G (eds) Current opinion in lipidology. Vol 1, no. 2: Genetics and molecular biology. Current Sciences, London, pp 110–115

Bazri S, Korn ED (1973) Single bilayer liposomes prepared without sonication. Biochim Biophys Acta 298:1015–1019

Breslow JL (1989) Familial disorders of high density lipoprotein metabolism. In: Scriver CR, Beaudet AL, Sly WS, Valle D (eds) The metabolic basis of inherited disease, 6th edn. McGraw-Hill Information Services, New York, pp 1251–1266

Brunzell JD (1989) Familial lipoprotein lipase deficiency and other causes of the chylomicronemia syndrome. In: Scriver CR, Beaudet AL, Sly WS, Valle D (eds) The metabolic basis of inherited disease, 6th edn. McGraw-Hill Information Services, New York, pp 1165–1180

Carlsson LA, Holmquist L (1985a) Evidence for the presence in human plasma of lecithin:cholesterol acyltransferase activity (β-LCAT) specifically esterifying free cholesterol of combined pre-β- and β-lipoproteinsin. Acta Med Scand 218:197–205

Carlsson LA, Holmquist L (1985b) Evidence for deficiency of high density lipoprotein lecithin:cholesterol acyltransferase activity (α-LCAT) in fish-eye disease. Acta Med Scand 218:189–196

Carlsson LA, Holmquist L, Assmann G (1987) Different substrate specificities of plasma lecithin:cholesterol acyltransferase in fish eye disease and Tangier disease. Acta Med Scand 222:283–289

DeBacker G, Hulstaerdt F, DeMunck K, Rosseneu M, Van Parijs L, Dramaix M (1986) Serum lipids and apolipoproteins in students whose parents suffered from a myocardial infarction. Am Heart J 112:478–484

Dobiasova M (1983) Lecithin:cholesterol acyltransferase and the regulation of endogenous cholesterol transport. Adv Lipid Res 20:107–144

Forte TM, Nichols AV, Krauss RM, Norum RA (1984) Familial apolipoprotein A-I and C-III deficiency: subclass distribution, composition and morphology of lipoproteins in a disorder associated with premature atherosclerosis. J Clin Invest 74:1601–1613

Franceschini G, Baio M, Calabresi L, Sirtor CR, Cheung MC (1990) Apolipoprotein A-I(Milano): partial familial lecithin cholesterol acyltransferase deficiency. Biochim Biophys Acta 1043:1–6

Frohlich J, McLeod R, Pritchard PH, Fesmire J, McConathy W (1988) Plasma lipoprotein abnormalities in heterozygotes for familial lecithin:cholesterol acyltransferase deficiency. Metabolism 37:3–8

Funke H, von Eckardstein A, Pritchard PH, Karas M, Albers JJ, Assmann G (1991) A frameshift mutation in the apo A-I gene causes corneal opacities, HDL deficiency and partial LCAT deficiency. J Clin Invest 87:375–380

Glomset JA (1968) The plasma lecithin:cholesterol acyltransferase reaction. J Lipid Res 9:155–163

Gordon D, Rifkind BM (1989) Current concepts: high density lipoproteins – the clinical implications of recent studies. N Engl J Med 321:1311–1315

Hunt SC, Hasstedt SJ, Kuida H, Stults BM, Hopkins PN, Williams RR (1989) Genetic heritability and common environmental components of resting and stressed blood pressures, lipids, and body mass index in Utah pedigrees and twins. Am J Epidemiol 129:625–638

Jonas A, von Eckardstein A, Kezdy K, Steinmetz A, Assmann G (1991) Structural and functional properties of reconstituted high density lipoprotein discs prepared with six apolipoprotein variants. J Lipid Res 32:90–100

Nichols WC, Dwulet F, Liepnieks J, Benson MD (1988) Variant apolipoprotein AI as a major constitutent of a human hereditary amyloid. Biochem Biophys Res Commun 156:762–768

Norum KR, Gjone E, Glomset JA (1989) Familial lecithin:cholesterol acyltransferase deficiency including fish-eye disease. In: Scriver CR, Beaudet AL, Sly WS, Valle D (eds) The metabolic basis of inherited disease, 6th edn. McGraw-Hill Information Services, New York, pp 1181–1194

Pometta D, Micheli H, Suenram A, Jornot C (1979) HDL lipids in close relatives of coronary heart disease patients: environmental and genetic influences. Atherosclerosis 34:419–429

Pritchard PH, McLeod R, Frohlich J, Park MC, Kudchodkar BJ, Lacko AG (1988) Lecithin cholesterol acyltransferase in familial HDL deficiency (Tangier disease). Biochim Biophys Acta 958:227–234

Utermann G, Schoenborn W, Langer KH, Dieter P (1972) Lipoproteins in LCAT-deficiency. Hum Genet 16:295–302

von Eckardstein A, Funke H, Henke A, Altland K, Benninghoven A, Assmann G (1989) Apolipoprotein A-I variants: naturally occurring substitutions of proline residues affect the plasma concentration of apolipoprotein A-I. J Clin Invest 84:1722–1730

von Eckardstein A, Funke H, Walter M, Atland K, Benninghoven A, Assmann G (1990) Structural analysis of apolipoprotein A-I variants: amino acid substitutions are nonrandomly distributed throughout the apolipoprotein A-I primary structure. J Biol Chem 265:8610–8617

Wilson DE, Emi M, Iverius PH et al. (1990) Phenotypic expression of heterozygous lipoprotein lipase deficiency in the extended pedigree of a proband heterozygous for a missense mutation. J Clin Invest 86:735–750

Chapter 16

Cholesterol-Lowering Clinical Trials: Where Do We Go From Here?

A. M. Gotto, Jr

Introduction

The contributors to this volume have raised many interesting ideas and proposals about the future direction of atherosclerosis research in the short and long term. The important information that has been gathered from clinical trials and the accomplishments of the late 1980s outline a highly impressive story. We now have firm evidence that the incidence of coronary heart disease (CHD) in the form of myocardial infarction and/or coronary death can be reduced – in both symptomatic and asymptomatic populations – by lipid modification; specifically, reduction in serum levels of low-density lipoprotein (LDL) cholesterol combined with increased serum levels of high-density lipoprotein (HDL) cholesterol. To choose where to go next in this research requires that we look at where we have been.

Some Previous Venues

The Lipid Research Clinics Coronary Primary Prevention Trial (LRC–CPPT) was mainly a test of the benefits of reducing LDL serum levels in asymptomatic subjects, and achieved excellent results (Fig. 16.1). All of the approximately 3800 subjects were hypercholesterolemic men, receiving the bile-acid seques-trant cholestyramine or placebo for an average of 7.4 years. In the cholestyramine group, mean falls of 8% and 12% in total cholesterol and LDL cholesterol relative to levels in the placebo group translated into a 19% lower

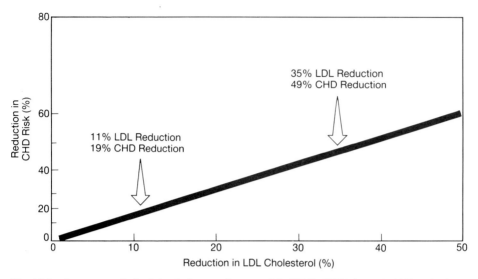

Fig. 16.1. Separate analysis of the cholestyramine arm of the LRC–CPPT showed a 19% reduction in CHD to be associated with each 11% decrement in LDL cholesterol. (Data from Lipid Research Clinics Program 1984.)

incidence of CHD (in this study, comprising definite CHD death and/or definite non-fatal myocardial infarction), also relative to placebo. When the cholestyramine group was considered separately, each 8% decrease in total cholesterol or 11% decrease in LDL cholesterol was accompanied by a 19% drop in CHD risk (Lipid Research Clinics Program 1984).

From the LRC–CPPT results was derived the 2:1 cholesterol-lowering ratio: namely, that a 1% reduction in total cholesterol is associated with a 2% decrease in coronary events. Richard Peto of Oxford University suggested that over a longer period a 1% reduction in total cholesterol would result in a 3% drop in the incidence of coronary events (personal communication 1987). In the Helsinki Heart Study, after only five years a 34% decrease in CHD incidence (again defined by cardiac death and non-fatal myocardial infarction) was seen in dyslipidemic men treated with the fibric-acid derivative gemfibrozil, in association with a 10% decrease in total cholesterol, an 11% decrease in LDL cholesterol, and an 11% increase in HDL cholesterol, as well as a 35% decrease in triglycerides, all values relative to placebo (Manninen et al. 1988). Proportional hazards analysis showed that, viewed separately, a 7% decrease in LDL, an 8% increase in HDL, and a 24% decrease in triglycerides would reduce CHD risk 15%, 23%, and 12%; their joint effect would be a 27% reduction. A 10% decrease in non-HDL cholesterol would lower CHD risk 22%. Nearly 3000 men completed the Helsinki Heart Study.

In both of these primary prevention trials, there was a lag of one or two years before a reduction in coronary events was observed. The subjects who benefited most from gemfibrozil treatment in the Helsinki Heart Study were those who had type IIB hyperlipidemia (Fig. 16.2), namely, an elevation of LDL cholesterol, an increase in triglycerides, and relatively low levels of HDL

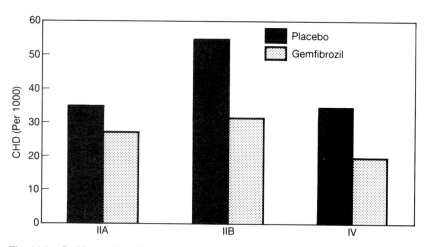

Fig. 16.2. Incidence of cardiac events by Fredrickson type of hyperlipidemia in the Helsinki Heart Study. Incidence rates standardized for smoking and hypertension were 28, 33, and 21 per 1000 in the gemfibrozil group and 35. 55, and 34 per 1000 in the placebo group for Fredrickson types IIA, IIB, and IV. (Adapted from Manninen et al. 1988.)

cholesterol. By contrast, the patients who benefited most from cholestyramine treatment in the LRC–CPPT were those who had elevated levels of LDL along with high levels of HDL (Lipid Research Clinics Program 1984). In neither trial did all-cause mortality significantly decrease. The drop in coronary deaths was offset by a rise in non-cardiovascular deaths, particularly in violent deaths (e.g. from traffic accidents or suicide).

In this author's opinion, in neither study was the power sufficient to demonstrate a difference in all-cause mortality. Rossouw, Lewis, and Rifkind (1990) included the LRC–CPPT and Helsinki trials among the four primary prevention studies assessed in their recent meta-analysis of cholesterol-lowering trials. Reducing serum cholesterol 10% across these four trials led to reductions of 25%, 12%, and 22% in the number of non-fatal, fatal, and all myocardial infarctions. Their analysis of eight major secondary-prevention trials (post-myocardial infarction) (Table 16.1) showed corresponding reductions of 19%, 12%, and 15% for each 10% mean reduction in total cholesterol. The reductions in mortality in the secondary-prevention trials (Table 16.2) were less dramatic than the reductions in myocardial infarction, but "indicated that there are more grounds for optimism in regard to both total mortality and non-cardiovascular mortality than was apparent in the primary prevention trials" (Rossouw et al. 1990, p. 1116). These authors recommended that a complete (fasting) lipoprotein analysis be performed in all patients with established CHD, and concluded that cholesterol lowering should be actively pursued in most CHD patients.

To clarify the relation of HDL cholesterol levels to CHD risk, Gordon et al. (1989) reassessed by multivariate analysis the data from several large epidemiological and randomized clinical trials. They found that each 1 mg/dl increment in HDL cholesterol meant a 2%–3% decrement in CHD risk. This consistent inverse relationship accords well with the results of the Helsinki Heart

Study (which study was not included in the Gordon analysis), namely, that a 1% increase in HDL was associated with 3% decrease in coronary events. Gordon et al. found HDL levels to be essentially unrelated to non-cardiovascular disease mortality.

Table 16.1. Rates of myocardial infarction in secondary prevention trials of cholesterol lowering

Trial	Non-fatal		Fatal		All	
	T	C	T	C (%)	T	C
Coronary Drug Project						
Clo	13.1	13.8	17.7	19.2	28.0	30.1
N	10.2	13.8[a]	18.1	19.2	25.6	30.1[b]
Newcastle (Clo)	12.3	18.2	10.2	17.4[b]	21.3	32.0[a]
Edinburgh (Clo)	7.1	11.2	9.7	9.5	15.4	19.6
Stockholm (Clo + N)	12.5	18.1	16.8	26.4[a]	25.8	36.2[b]
Oslo (diet)	11.7	15.0	18.0	24.3	29.6	39.3[b]
Medical Research Council						
Low-fat diet	16.3	14.7	8.1	9.3	24.4	24.0
Soybean oil	10.1	13.4	12.6	12.9	22.6	26.3
Sum, observed less expected events	−63[c]		49[a]		−98[c]	
Odds ratio (95% CI)	0.75 (0.65–0.85)		0.84 (0.74–0.94)		0.78 (0.70–0.86)	

Source: Adapted from Rossouw et al. (1990).
Abreviations: C, control; Clo, clofibrate; N, nicotinic acid; T, treatment.

[a] $P < 0.01$.
[b] $P < 0.05$.
[c] $P < 0.001$.

Table 16.2. Mortality rates in secondary prevention trials of cholesterol lowering

Trial	Non-fatal		Fatal		All	
	T	C	T	C (%)	T	C
Coronary Drug Project						
Clo	21.8	22.7	2.6	1.9	25.5	25.4
N	21.3	22.7	2.7	1.9	24.4	25.4
Newcastle (Clo)	NA	NA	NA	NA	NA	NA
Edinburgh (Clo)	NA	NA	NA	NA	NA	NA
Stockholm (Clo + N)	19.4	27.2	2.5	2.5	21.9	29.7
Oslo (diet)	18.4	25.2	1.5	1.5	19.9	26.7
Medical Research Council						
Low-fat diet	NA	NA	NA	NA	16.3	18.6
Soybean oil	13.6	12.9	0.1	3.1	14.1	16.0
Sum, observed less expected events	−34[a]		9		−29	
Odds ratio (95% CI)	0.88 (0.77–0.99)		1.30 (0.93–1.83)		0.91 (0.82–1.02)	

Source: Adapted from Rossouw et al. (1990).
Abreviations: C, control; Clo, clofibrate; N, nicotinic acid; T, treatment.

[a] $P < 0.05$.

Table 16.3.   Lipid changes in surgical arm of POSCH at five years

Lipid	Mean (±SE) percentage change[a]	P value
Total cholesterol	−23.3 (±1.0)	<0.0001
LDL cholesterol	−37.7 (±1.2)	<0.0001
HDL cholesterol	+4.3 (±1.8)	0.02
VLDL cholesterol	+18.3 (±7.5)	0.02
Triglycerides	+19.8 (±6.5)	0.003
HDL cholesterol:total cholesterol	+37.8 (±2.8)	<0.0001
HDL cholesterol:LDL cholesterol	+71.8 (±4.3)	<0.0001

Data from Buchwald et al. (1990).
[a] Compared with control.

Also of interest are the results recently reported by the Program on the Surgical Control of the Hyperlipidemias (POSCH), a secondary-prevention trial that examined the effects of partial ileal bypass on lipid values and on CHD and overall mortality rates (Buchwald et al. 1990). The subjects were over 800 myocardial infarction survivors, half randomized to a control arm, followed up for a mean of 9.7 years. The bypass procedure results in a loss of bile acids and presumably would be similar in effect to bile-acid resins. Through the interference with enterohepatic circulation, the level of cholesterol in hepatocytes would be decreased, resulting in an increase in LDL-receptor activity. This trial included some women (9%).

The mean lipid changes at five years in the surgical arm compared with the control arm of the POSCH trial are shown in Table 16.3. The reductions in total and LDL cholesterol – 23% and 38% – are dramatic. HDL cholesterol was increased, but so were very-low-density lipoprotein (VLDL) cholesterol and triglycerides. Hypertriglyceridemia is known to be associated with malabsorption in the small bowel. The HDL-to-total cholesterol ratio improved by 38%, and the HDL-to-LDL ratio improved by 72% relative to controls. The lipid changes were maintained over the ten years of the POSCH follow-up in all of the surgery-group patients except a few who had the ileal bypass taken out.

Also highly significant in POSCH was the 35% reduction in the combined endpoint of CHD death and confirmed non-fatal myocardial infarction among patients who underwent partial ileal bypass ($P<0.001$), nearly identical to the 34% reduced CHD risk in the Helsinki Heart Study. As in the LRC–CPPT and the Helsinki Heart Study, about two years elapsed before changes in endpoints were seen. After that length of time, combined endpoint curves diverged significantly between the surgical and non-surgical groups (Fig. 16.3). Also significantly reduced in the surgery group were the rates of coronary artery bypass grafting (reduced by close to two-thirds) and percutaneous transluminal angioplasty (reduced by about one-half). Coronary arteriographic analyses were carried out at three, five, seven and ten years. At ten years, 85% of the non-surgical group showed worsening of the degree of coronary stenosis, compared with 55% in the surgery group. No change was documented in 11% of the non-surgical group and 39% of the surgery group, whereas 4% in the non-surgical group showed improvement, versus 6% in the surgery group.

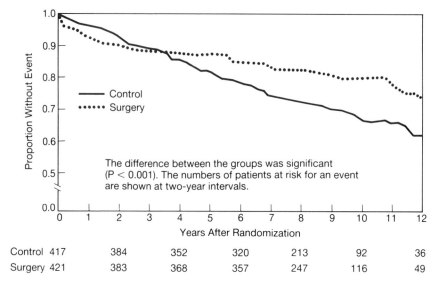

Fig. 16.3. Confirmed myocardial infarction and death due to atherosclerotic CHD as a combined endpoint ("event") in the POSCH trial. (Adapted from Buchwald et al. 1990.)

There were 62 deaths in the non-surgical group, 44 due to CHD, and 49 deaths in the surgery group, 32 due to CHD, but these reductions in mortality in the bypass group were not significant. On subgroup analysis, improved overall survival was seen among the patients in whom the ejection fraction was 50% or higher. This type of post hoc analysis cannot be given any credence from a strict statistical standpoint; nonetheless, it is reasonable to think that in those patients who had sustained a great deal of myocardial damage at the time of infarct, the quantity of damage was the major determinant of survival.

The most common side effect of partial ileal bypass in the POSCH study was diarrhea; also encountered were kidney stones, gallstones, and intestinal obstruction. There were no immediate postoperative deaths in hospital.

Several of the important recent drug trials have also used angiographic documentation. In the Cholesterol-Lowering Atherosclerosis Study (CLAS), middle-aged men who had undergone coronary bypass surgery were randomized to treatment by (1) diet and placebo, or (2) diet, the bile-acid resin colestipol, and nicotinic acid. Treatment doses received averaged about 30 g of colestipol and about 4 g of nicotinic acid per day throughout the study.

Of the 188 men enrolled, 162 completed baseline and two-year angiograms (CLAS I, Blankenhorn et al. 1987); 103 (of 138 who entered the second phase) continued to complete four-year angiograms (CLAS II, Cashin-Hemphill et al. 1990). After two years, total cholesterol in the drug group was decreased by 27%, LDL was decreased by 43%, and HDL was increased by 37% from baseline. The mean triglyceride value was also decreased, by 22%. These changes were maintained through the four years on trial (Table 16.4). LDL cholesterol was under 100 mg/dl at the end of the study, considerably lower than the 130 mg/dl boundary set by the National Cholesterol Education Program adult treatment guidelines in the United States.

Table 16.4.   Mean percentage lipid changes[a] in CLAS at two and four years

Lipid	CLAS I (2 years)		CLAS II (4 years)		P value[b]
	D	Pl	D	Pl	
Total cholesterol	−27	−4	−25	−6	<0.001
LDL cholesterol	−43	−4	−40	−6	<0.001
HDL cholesterol	+37	+2	+37	+2	<0.001
Triglycerides	−22	−5	−18	−5	<0.03

Data from Cashin-Hemphill et al. (1990).
Abbreviations: D, drug group; Pl, placebo group.
[a] From baseline.
[b] For group differences at two and four years.

The quantitative coronary angiography showed that at both two and four years the drug-treated group had significantly less progression of disease and fewer new lesions in the bypass grafts and native vessels than the placebo group. After four years, the placebo group showed an average of 2.0 native artery lesions that progressed per patient, versus 0.9 in the drug group ($P = 0.0002$), and 40% and 14%, respectively, had new lesions in native arteries ($P = 0.001$), whereas 38% versus 16% had new lesions in the grafts ($P = 0.006$). The CLAS researchers also evaluated changes in global coronary score – i.e. an average scoring of bypass grafts and native vessels – since coronary artery blockage in a given patient may be worse in one vessel but improved in another. After two years, 16% of the drug recipients showed evidence of regression by global scoring, compared with 2% of placebo recipients ($P = 0.002$); after four years, the figures were 18% and 7%.

Thus, aggressive lipid-lowering pharmacotherapy over a four-year period, associated with dramatic decreases in LDL, increases in HDL, and decreases in triglycerides, resulted in significantly less objective disease in both bypass grafts and native vessels in the CLAS trials. It should be noted that the doses of drugs used were very large, and many patients would be unable to tolerate them.

According to CLAS data, one of the best indicators of lack of coronary lesion progression is a high proportion of apolipoprotein C-III in HDL particles (Blankenhorn et al. 1990a). Such a value would indicate a rapid rate of catabolism of triglyceride-rich lipoproteins. In another CLAS analysis, of data from the placebo arm of the study, the development of new atherosclerotic lesions was statistically associated with total fat consumption. The placebo recipients who consumed no more than 23% of their total calories as fat showed the fewest new lesions, whereas those with fat intake of 34% or higher had the highest rate of new-lesion formation (Blankenhorn et al. 1990b). These findings argue strongly for the efficacy of a low-fat diet. Three fatty acids emerged in this analysis as significant risk factors for the development of new lesions, namely, lauric, oleic and linoleic acids.

The recent Lifestyle Heart Trial of Ornish et al. (1990) reported a favourable result on coronary stenosis with a low-fat vegetarian diet coupled with exercise, yoga and meditation. After one year, the special-intervention group showed significant overall regression of coronary lesions: 18 (82%) of the 22 patients had

Table 16.5. Mean percentage lipid changes[a] in FATS at 2½ years

Treatment group	LDL	HDL
Conventional therapy (with placebo; colestipol if LDL elevated)	−7	+5
Lovastatin + colestipol	−46	+15
Nicotinic acid + colestipol	−32	+43

Data from Brown et al. (1990).
[a] From baseline.

changes in the direction of regression, and the average percentage diameter stenosis among the 22 decreased from 40% to 38%. In contrast, 10 (53%) of the 19 patients randomized to control had on average lesion change toward progression; the percentage diameter stenosis increased from 43% to 46% among the controls. Five of Ornish's 41 patients were women.

In the Familial Atherosclerosis Treatment Study (FATS) recently published by Brown et al. (1990), the entry criteria included an apolipoprotein B-100 concentration greater than 125 mg/dl. This study was limited to men with a family history of CHD who had stenosis of at least 50% in one coronary artery or 30% or more in three coronary arteries, as determined by examining proximal segments using a quantitative method. In general, the patients had severe CHD. Stenosis was re-evaluated at 30 months by the same method.

The 146 men enrolled in FATS were given dietary counselling and were randomly assigned to receive: (1) placebo, with colestipol given if the LDL level remained above the 90th percentile by age on conventional therapy; (2) lovastatin (40 mg/day) plus colestipol (30 g/day); or (3) nicotinic acid (4 g/day) plus colestipol (30 g/day). If the LDL level did not fall below 120 mg/dl, lovastatin or nicotinic acid was increased (to 80 mg/day or 6 g/day); 120 men completed the study.

The drug interventions yielded dramatic changes in the plasma lipid and apolipoprotein concentrations. The greatest reduction of LDL was in the lovastatin–colestipol group, and the greatest increase in HDL was in the nicotinic acid–colestipol recipients (Table 16.5). The reductions in apolipoprotein B-100 more or less paralleled those of LDL cholesterol.

Further, the intensive lipid-lowering therapy in FATS reduced the frequency of progression of coronary lesions and increased the frequency of regression (Table 16.6). On the basis of previous experience, it was expected that the controls would show evidence of progression of stenosis of about 1% per year. This is what was observed, whereas each of the drug-treated groups showed regression, a result that was statistically significant (Fig. 16.4). Fewer drug recipients experienced cardiac events (death, myocardial infarction, revascularization for worsening symptoms): 3 of 46 randomized to lovastatin–colestipol and 2 of 48 randomized to nicotinic acid–colestipol, compared with 10 of 52 placebo recipients.

Another recently reported angiographic drug trial is that conducted at the University of California, San Francisco (UCSF) in 72 patients, 41 of them women, with familial hypercholesterolemia (Kane et al. 1990). The patients

Table 16.6. Arteriographic changes in FATS at 2½ years

Treatment group	Patients with progression as only change[a] (%)	Patients with regression as only change[b] (%)
Conventional therapy (with placebo; colestipol if LDL elevated)	46	11
Lovastatin + colestipol	21	32
Nicotinic acid + colestipol	25	39

Data from the Familial Atherosclerosis Treatment Study, Brown et al. (1990).
[a] Among the worst lesions measured in each of a standard set of nine proximal coronary segments, at least one worsened by at least 10% stenosis, and none improved by that amount.
[b] Converse of the above. $P < 0.005$.

were randomized to diet (alone or with low-dose colestipol) or diet plus aggressive LDL-lowering drug therapy (colestipol, nicotinic acid, lovastatin – in various combinations). Average time on study was about two years. This study showed significant lipid improvements with drug therapy: -38% in LDL cholesterol (the same in each sex), -19% in triglycerides (-11% in men, -25% in women), and $+28\%$ in HDL cholesterol ($+29\%$ in men, $+27\%$ in women). The P values for both sexes against control were <0.001, 0.05, and 0.03, respectively.

The average change in mean percentage stenosis in the UCSF study was -1.53% in the drug-treated group, indicative of lesion regression, and $+0.80\%$ in the control group, indicative of lesion progression ($P = 0.039$). The significance was retained when angiographic results in women were looked at separately (-2.06% versus $+1.07\%$, $P = 0.05$), but not retained in men (-0.88% versus $+0.41\%$, $P = 0.42$).

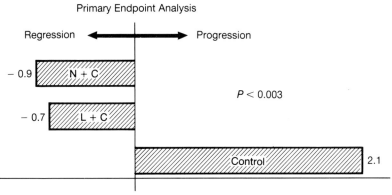

Fig. 16.4. Mean percentage change in stenosis by FATS treatment group after 2½ years on study. Change is from a baseline of 34%, representing the average percentage of stenosis caused by the worst lesion in each of nine proximal segments. The P value is for the trend. (Figure courtesy of Greg Brown and colleagues, University of Washington School of Medicine.)

Future Directions

In the CLAS results, the benefits of lipid lowering were seen across the entire range of baseline lipid values, namely, 185–350 mg/dl (Blankenhorn et al. 1987). In future trials, it will be important to determine what relative changes are needed for greatest benefit in total cholesterol and the LDL and HDL parameters so critical to the success of trials to date. Definition of groups most likely to show regression or progression would be of extraordinary value.

In FATS, the most advanced lesions were the most likely to show regression (Brown et al. 1990). In addition, it is hoped that researchers will be able to identify at which anatomical sites subsequent coronary events are most likely. It has been previously suggested that thrombosis is more likely to occur at sites with a smaller degree of stenosis, which is counter to the impressions in POSCH (H. Buchwald, personal communication, 1990).

Another entrée to more useful data will be to continue to expand lipid-lowering trials to new patient groups. The UCSF findings in women are of great interest in this regard.

These suggestions are but a few guideposts for future clinical trials, midway between the many questions that pertain to the areas of basic science and practical application in this field.

References

Blankenhorn DH, Nessim SA, Johnson RL et al. (1987) Beneficial effects of combined colestipol–niacin therapy on coronary atherosclerosis and coronary venous bypass grafts. JAMA 257:3233–3240

Blankenhorn DH, Alaupovic P, Wickham W, Chin HP, Azen SP (1990a) Prediction of angiographic change in native human coronary arteries and aortocoronary bypass grafts: lipid and nonlipid factors. Circulation 81:694–696

Blankenhorn DH, Johnson RL, Mack WJ, El Zein HA, Vailas LI (1990b) The influence of diet on the appearance of new lesions in human coronary arteries. JAMA 263:1646–1652

Brown G, Albers JJ, Fisher LD et al. (1990) Regression of coronary artery disease as a result of intensive lipid-lowering therapy in men with high levels of apolipoprotein B. N Engl J Med 323:1289–1298

Buchwald H, Varco R, Matts J et al. (1990) Effect of partial ileal bypass surgery on mortality and morbidity from coronary heart disease in patients with hypercholesterolemia. N Engl J Med 323:946–955

Cashin-Hemphill L, Mack WJ, Pogoda JM et al. (1990) Beneficial effects of colestipol–niacin on coronary atherosclerosis: a 4-year follow-up. JAMA 264:3013–3017

Gordon D, Probstfield J, Garrison R et al. (1989) High density lipoprotein cholesterol and cardiovascular disease: four prospective American studies: Circulation 79:8–15

Kane JP, Malloy MJ, Ports TA, Phillips NR, Diehl JC, Havel RJ (1990) Regression of coronary atherosclerosis during treatment of familial hypercholesterolemia with combined drug regimens. JAMA 264:3007–3012

Lipid Research Clinics Program (1984) The Lipid Research Clinics Coronary Primary Prevention Trial results. I. Reduction in incidence of coronary heart disease. II. The relationship of reduction in incidence of coronary heart disease to cholesterol lowering. JAMA 251:641–651

Manninen V, Elo MO, Frick MH et al. (1988) Lipid alterations and decline in the incidence of coronary heart disease in the Helsinki Heart Study. JAMA 260:641–651

Ornish D, Brown SE, Scherwitz SW et al. (1990) Can lifestyle changes reverse coronary heart disease? The Lifestyle Heart Trial. Lancet 336:129–133

Rossouw JE, Lewis B, Rifkind BM (1990) The value of lowering cholesterol after myocardial infarction. N Engl J Med 323:1112–1119

Subject Index